Parasites cause many important diseases in humans and domestic animals, malaria being an example. Parasites have evolved to exploit hosts' bodies whereas hosts have evolved immune systems to control infections. Host–parasite interactions therefore provide fascinating examples of evolutionary 'arms-races' in which the immune system plays a key role. Modern research in immunoparasitology is directed towards understanding and exploiting the capacity to develop effective anti-parasite immunity. By concentrating on selected infections where research has made significant progress, *Immunity to Parasites* provides a clear account of how host immune responses operate and how parasites can evade immunity. The experimental basis of this research is emphasized throughout. This completely updated second edition includes an expanded section on anti-parasite vaccines. The text is aimed at undergraduates and postgraduates with interests in either parasitology or immunology and provides introductory sections on these topics to lead the reader into the later chapters.

Immunity to parasites

How parasitic infections are controlled

SECOND EDITION

DEREK WAKELIN

D.Sc., F.R.C. Path., Professor of Zoology, University of Nottingham

PUBLISHED BY THE PRESS SYNDICATE OF THE UNIVERSITY OF CAMBRIDGE
The Pitt Building, Trumpington Street, Cambridge CB2 1RP, United Kingdom

CAMBRIDGE UNIVERSITY PRESS
The Edinburgh Building, Cambridge CB2 2RU, UK http://www.cup.cam.ac.uk
40 West 20th Street, New York, NY 10011-4211, USA http://www.cup.org
10 Stamford Road, Oakleigh, Melbourne 3166, Australia

First published 1984 by Edward Arnold
Second edition published by Cambridge University Press 1996
Reprinted 1998

Typeset in Monotype Garamond 11/13pt, in QuarkXPress™ [SE]

A catalogue record for this book is available from the British Library

Library of Congress Cataloguing in Publication data

Wakelin, Derek.
    Immunity to parasites : how parasitic infections are controlled /
Derek Wakelin.
        p.   cm.
    Includes bibliographical references.
    ISBN 0 521 56245 7 (hardcover). – ISBN 0 521 43635 4 (pbk.)
    1. Parasitic diseases – Immunological aspects.   I. Title.
QR201. P27W35   1996
616.9′6079–dc20   95-48213 CIP

ISBN 0 521 56245 7 hardback
ISBN 0 521 43635 4 paperback

Transferred to digital printing 2001

Immunity to parasites

How parasitic infections are controlled

For P. C. W.

Contents

Preface to the second edition

In the decade since the first edition of *Immunity to Parasites* there has been an explosive growth of interest and activity in immunoparasitology, fuelled by developments in two areas of biological science. The application and exploitation of molecular techniques has revolutionized our concepts of parasites as triggers and targets for immune responses, providing detailed knowledge of parasite antigens. The extraordinary progress in immunology over this period has provided a level of understanding of the induction and regulation of host responses that was almost inconceivable in the early 1980s. The combination of the two has allowed precise and imaginative experimental investigation of host–parasite relationships and the extension of such analyses to parasitic infections in human populations.

The consequences of this growth have made it more difficult to provide a selective picture of immunoparasitology. The choice of parasites for the first edition was much simpler than it is now. In 1984 relatively few host–parasite systems were well studied or understood in any detail. Now we have detailed insights into many more, obtained from both clinical investigations and experimental systems. All of these could be chosen to illustrate important principles about immunity to parasites, but to do so would alter both the size and the approach of this particular text. I am conscious that any decision to be selective can be criticized, and those who might justifiably expect to see accounts of parasites such as *Eimeria, Entamoeba, Giardia, Trypanosoma cruzi, Toxoplasma* and the tapeworms, in which there have been substantial advances, will be disappointed. I have considered it most useful to continue with the groups of parasites covered in the first edition, both to provide continuity and because these do still provide good examples from which to draw more general conclusions. The last ten years have also seen an enormous increase in immunoparasitological literature and in the rate at which new information appears. It is therefore difficult to decide how much and how recent should be the information presented in a book of this kind, given that many statements made now will date quite quickly. I have compromised by trying to select information that makes it possible to understand the biological bases that underlie the immunological aspects of the host–parasite relationship, whilst conveying some of the excitement that the rapid growth in fundamental understanding has brought to the subject. One area in which this excitement is most intense is the progress towards effective anti-parasite vaccines, and this is the one section that has been substantially increased.

Nottingham, 1995   D.W.

Acknowledgements

I am grateful to my colleagues in the Department of Life Science at Nottingham, Dr Jerzy Behnke and Dr David Pritchard, for advice and helpful suggestions. Particular thanks go to Professor Jennie Blackwell, Dr Paul Hagan, Professor Stephen Phillips and Professor Keith Vickerman FRS, who very kindly read the draft chapters on *Leishmania*, Schistosomes, Malaria and Trypanosomes; any errors that remain in these are entirely my responsibility.

All drawings that are not original have been adapted from the sources acknowledged in the legends. I am greatly indebted to the following for permission to reproduce their photographs: Dr R. Nussenzweig and Academic Press, Fig. 4.3; Dr L.H. Bannister and *Parasitology* (Cambridge University Press), Fig. 4.6; Professor R.E. Sinden, Fig. 4.8; Professor Dr H. Melhorn, Figs. 4.10, 4.15, 6.3, 9.1a; Professor J. Alexander and Academic Press, Fig. 4.17a; Dr S.L. Croft, Figs. 4.17b, 8.8; Professor J. Alexander and *Transactions of the Royal Society of Tropical Medicine and Hygiene*, Fig. 4.17c; Dr L. Tetley and Professor K. Vickerman, Figs. 5.2, 5.4; Dr D.J. McLaren, Fig. 6.7; Dr D.J. McLaren and *Parasitology* (Cambridge University Press), Fig. 6.8a; Dr D.J. McLaren and John Wiley, Figs. 6.8b,c; Professor Y. Takahashi and *Parasitology Research*, Fig. 7.6; Dr R.B. Crandall, Fig. 8.8; Dr P. Willadsen and P. Tracey-Patte, Fig. 9.1b.

Preface to the first edition

Parasitic animals constitute one of the major causes of the infectious diseases which affect man and his domestic animals. The cumulative effects of parasitism in terms of mortality, chronic disease and economic loss are incalculably great and undoubtedly play an important part in limiting the social and economic development of many countries in sub-tropical and tropical regions of the world. It should not be thought that man and domestic stock are in some way uniquely vulnerable to parasitic infections. All species are subject to infection by a characteristic assemblage of parasitic animals adapted to them, and these relationships have arisen as a result of long evolutionary development. Exploitation of the environments provided by the bodies of living animals is, in essence, no different from the exploitation of the environments provided by the seas, the land and fresh waters, and has been a major line of evolution followed by species from many phyla. The success of this way of life is reflected in the extent to which animals of every phylum are parasitized and in the stability and antiquity of many host–parasite relationships. Stability is often reflected in minimal inconvenience to the life and survival of the host and is dependent upon a balanced relationship not only between individual hosts and individual parasites, but also between populations of each. Thus, although man and domestic stock are not exceptional in their susceptibility to parasites, both may live under conditions that predispose them to heavy and continuous infection. In consequence the balance of the host–parasite relationship is disturbed and the host may suffer severely from parasite disease.

The almost universal occurrence of parasites has exerted a major influence upon the evolution of the protective responses which allow hosts to regulate the levels of infections to which they are subject. All hosts have the capacity to respond adaptively to the presence of parasites and to exert deleterious effects upon their growth and survival. All parasites, therefore, have had to evolve strategies by which they can circumvent these effects. In higher vertebrates this adaptive response is very largely, though not exclusively, the property of the immune system, and immunologically based, anti-parasite responses represent a major survival strategy. The study of these responses has obvious relevance to human and veterinary medicine, particularly in the search for more effective means of control, and in recent years has attracted a great deal of research interest from both parasitologists and immunologists. Warren has differentiated between 'parasite-immunologists', trained in immunology, and 'immuno-parasitologists', trained in parasitology. Both, however, are required to make biologically meaningful studies of immune responses to parasites. Each complements the other, and the growth of their research field, which I prefer to call immunoparasitology, is a satisfying example of mutualism in action.

Immunoparasitology has an obvious applied significance. It is also an intriguing area of study in its own right, throwing light upon fundamental aspects of protective immune responses and illuminating the intricacies of host–parasite interactions. The results of immunoparasitological research are now published in a wide variety of journals and have been summarized in a number of excellent texts and reviews. Few of these sources are entirely appropriate for the undergraduate or graduate biologist seeking an insight into the immunological aspects of parasitology, or for the immunologist wishing to put immunoparasitology into a biological context. The aim of this volume is to meet these needs by concentrating on selected host–parasite relationships in which immunologically-orientated research has made significant progress, both in elucidating the nature of host-protective responses and in clarifying the reciprocal interactions between these responses and parasite adaptations. The introductory chapters are intended to help readers of differing backgrounds into the subject area. That on immunology is designed primarily to introduce aspects of the immune response which are relevant to immunoparasitology. Chapters 4 to 9 then consider a limited number of examples chosen to illustrate the diversity of organisms and adaptations with which the immune response must deal. The emphasis in these chapters falls upon experimental studies using laboratory model systems, in which particular problems can be analysed in great detail without the many complicating factors that bedevil analysis of protective responses in man and domestic animals. Nevertheless, the ultimate goal of experimental studies is to provide data applicable to the understanding and the control of human and animal disease, and the final chapter considers the ways in which such studies are contributing to this objective.

This is not intended to be a comprehensive account of immunoparasitology and, inevitably, much that can rightly be considered significant has had to be omitted. It is very much a personal overview; nevertheless, it will, I hope, be useful to a wide variety of readers and contribute a little to the growth and development of a fascinating science.

Nottingham, 1984   D.W.

1

Parasites and parasitism

The host as an environment

1.1   Introduction

In many older textbooks of zoology it was customary to treat parasite species as though they were in some way quite separate from free-living animals. The term 'degenerate', which was used to describe the apparent simplifications in structure shown by parasites when compared with their free-living relatives, also carried with it a whiff of moral disapproval for their way of life! Today the pendulum has swung almost to the other extreme and parasitism is sometimes represented as nothing more than another form of environmental exploitation. It is indeed a remarkably successful exploitation and one that has been a major ine of evolutionary development in several phyla. The success of this way of life is attested by the ubiquity of parasites, in hosts of every phylum, and in the long-term stability of many host–parasite relationships. In essence, of course, it is true that exploitation of the host environment is very similar to the exploitation of any other environment, but there is one very important difference. Unlike the environments of free-living animals, the environment provided by a host can respond adaptively to the presence of a parasite. It is this difference, the adaptive interaction of host and parasite, each concerned with its own evolutionary survival, that distinguishes parasitism from other modes of life.

1.2   What is a parasite?

Despite the distinctive feature outlined above it is almost impossible to define parasitism in terms that completely exclude related modes of interspecific association such as commensalism and mutualism. In the host–parasite relationship the parasite is undoubtedly the beneficiary, but the precise requirements for parasite survival and the extent to which the host may suffer from the association are extremely variable. For the purposes of this book, which will be concerned almost exclusively with endoparasites in warm-blooded hosts (Table 1.1), it is useful to apply some ecological concepts in describing the essential characteristics of the parasitic way of life.

For parasites the host is the total environment. Larval and other reproductive stages may live in the outside world for longer or shorter periods but this represents merely a necessary phase in the movement from host to host. Particular parasites occupy particular niches in the major habitats provided by the host environment and are adapted to the conditions present in those niches in exactly the same way as free-living organisms are adapted to their environments. Although the environment is wholly biotic in origin, as it is provided by a living organism, it is still possible to define each niche by what are essentially abiotic factors such as pH, oxygen tension, redox potential and nutrient availability, as well as truly biotic factors such as other parasites and resident microorganisms. The ecological analogies break down, however, when one considers that the environment provided by the host is not passive, but can react adaptively to the presence of the parasite. Thus even in an environment to which they are perfectly adapted, parasites are faced by a variety of potentially destructive factors never experienced by free-living species, for example antibodies, complement, cytotoxic cells, lysosomal enzymes, and toxic metabolites as well as predatory phagocytic cells. The ability of the parasite to evade or resist these adaptive responses ultimately determines the ability of the parasite to survive and reproduce.

An important consequence of the endoparasitic condition is that the parasite is cut off from direct experience of the external world. Some parasites still rely on external changes in temperature and day length, which they detect indirectly through the hormonal and other changes in the host, to control their own developmental processes. Others (e.g. intestinal nematodes of sheep) use their direct experience of these changes, whilst they are in the external world as larval stages, to regulate their development after entry into the host. The majority coordinate their growth, development and reproduction by responding to factors present within the host environment that have little or no relationship to the outside world. Such adaptations have clear selective advantage in preventing developmental changes from taking place in the absence of suitable hosts.

Table 1.1 *Table of major parasites referred to in the text*

Classification	Genus	Position in host	Transmission	Size
Protozoa				
Mastigophora	*Trypanosoma*	Extracellular/Blood	Bite of tsetse fly	15–25 μm
	Leishmania	Intracellular/Macrophage	Bite of sandfly	2–5 μm
Apicomplexa	*Plasmodium*	Intracellular/RBC	Bite of mosquito	2–20 μm
Platyhelminthes				
Digenea	*Schistosoma*	Blood vessels (adults)	Skin penetration by larvae	10–30 mm (♀)
Nematoda				
Strongylida	Hookworms, *Haemonchus*, *Nippostrongylus*, *Trichostrongylus*, *Heligmosomoides*	Intestinal lumen	Oral ingestion or skin penetration by larvae	10 mm (♀)
	Dictyocaulus	Lungs		10 cm (♀)
Trichinelloidea	*Trichinella*	Intracellular/Gut epithelium	Oral ingestion of larvae	2–3 mm (♀)
Filaroidea	*Wuchereria*, *Brugia*	Lymphatics (adults)	Bite of mosquito	100 mm (♀)
	Onchocerca	Subcutaneous tissue (adults)	Bite of *Simulium*	500 mm (♀)
Arthropoda				
Acarina	'Ticks'	Ectoparasitic on skin	Direct host contact	5–15 mm (♀)

1.3   Parasites and parasite life cycles

1.3.1   Parasites

Animals that live as parasites occur in almost every phylum and animals of every phylum are subject to parasitic infection; the total number of parasite species is therefore enormous. Even if consideration is restricted only to parasites affecting humans and domestic animals, the number of species known is very large, yet for the majority, little or nothing is known of immunological aspects of their relationships with their hosts. Our present understanding of immunoparasitology comes from studies made on relatively few species and we have to assume that this understanding will prove to have general applicability. Table 1.1 includes some of the parasites that have been studied in detail, either because they are of clinical or veterinary significance, or because thay provide useful experimental models. The genera listed belong to the four major parasitic groups – Protozoa, Platyhelminthes, Nematoda and Arthropoda – and cover a wide size range, from very small intracellular parasites to very large extracellular worms, which may be several thousand times larger. It is useful to divide the groups into two major categories – *microparasites* (protozoans) and *macroparasites* (platyhelminths, nematodes and arthopods) – not only in terms of their size, but also in terms of their ability to increase numbers by replication within the host. Protozoans, together with the other major microparasites, viruses and bacteria, replicate within the host, and levels of infection can rise rapidly even after a single infection event, which, theoretically at least, need involve only a single organism. In contrast, the majority of macroparasites cannot replicate within the host, and levels of infection are determined by the number of infection events and the number of infective stages acquired. This biological distinction between micro- and macroparasites is therefore of fundamental importance in both epidemiology and immunoparasitology.

1.3.2   Life cycles

There is an almost infinite variety of ways in which parasites reach their hosts and of life cycles directed to this end, but it is possible to reduce this variety to a number of basic patterns.

(*a*) The parasite is never exposed to the external world and completes its development and reproduction in a single host. Transmission between hosts is normally achieved by predation or scavenging (Fig. 1.1a). Relatively few parasites show this pattern.

(*b*) The parasite is never exposed to the external world, but its development cycle takes place in two or more host species. The species in which the parasite reaches sexual maturity is known as the *final* or *definitive* host: that in which larval, juvenile or non-sexual stages develop is known as the *intermediate* host. Where one host transmits the parasite directly to another it is also referred to as a *vector* and this term is applied particularly to arthropods (Fig. 1.1b).

(*c*) The parasite is exposed to the external world for varying periods of time, but does not have active, free-living stages. The infective forms are contained within protective structures such as cysts (Protozoa) or egg shells (Nematoda). The life cycle may be *direct* in that only one host species is involved, or *indirect*, involving intermediate and final hosts of different species (Fig. 1.1c).

(*d*) The parasite is exposed to the external world as an active, free-living form during its development and transmission between hosts. Re-entry into the host may be a passive process, i.e. by ingestion, or an active process by penetration. As before, the cycle may be direct or indirect (Fig. 1.1d).

In all patterns, certain phases in the life cycle act as the infective stages and these may show well-defined morphological or physiological adaptations for this function. These are particularly evident in the skin-penetrating larvae of helminths, for example the cercariae of schistosomes, which possess suckers for adhesion and a battery of glandular structures that release histolytic enzymes. Characteristic of all infective stages is the capacity to recognize the host environment. In protozoan infective stages such as the malarial sporozoite, this capacity operates at a molecular level and involves complementarity between molecules present on the parasite and molecules expressed on host cell membranes. In the infective stages of helminths, particularly those that enter the host in a protective sheath, cyst or eggshell, recognition of the host involves response to certain physico-chemical stimuli that act as triggers to initiate escape from the protective layers and recommencement of growth and development. Clearly, despite numerous adaptations to ensure that it is the correct host species that is invaded, there is a large random element in the process. Host contact may or may not be made. If made, infection may fail because the host is unsuitable. The reasons underlying failures of the latter kind can be very instructive because, by throwing light upon the phenomenon of natural resistance or natural insusceptibility to infection, they reveal much about the characteristics of the susceptible host.

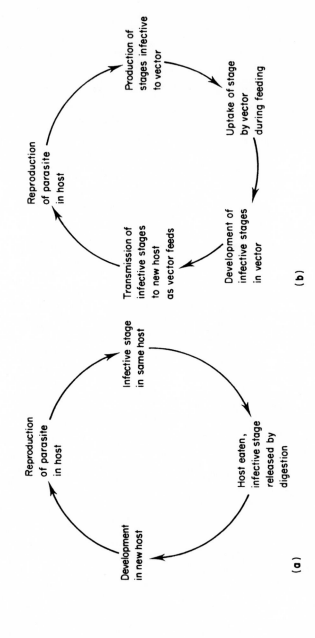

Fig. 1.1 Life cycles of endoparasites.
(a) Direct cycle, parasite never exposed to the outside world, transmission by carnivory, e.g. *Trichinella* (Nematoda).
(b) Indirect cycle, parasite never exposed to the outside world, transmission by arthropod vector, e.g. *Plasmodium* (Protozoa), *Wuchereria* (Nematoda).

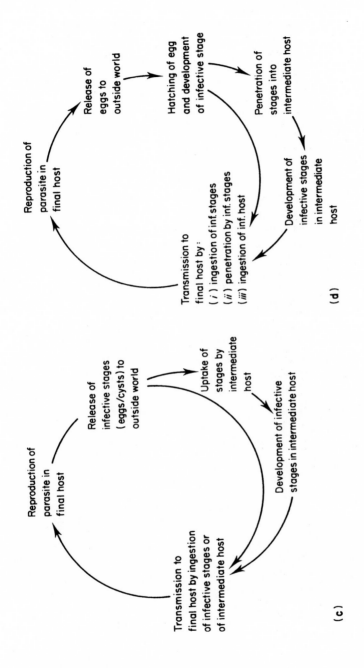

Fig. 1.1 (cont.)

(c) Direct and indirect cycles, parasite exposed to the outside world but not as an active free-living stage. Transmission direct, e.g. *Toxoplasma* (Protozoa), *Ascaris* (Nematoda), or indirect via an intermediate host, e.g. tapeworms (Platyhelminthes).

(d) Direct and indirect cycles, parasite exposed to the outside world as free-living stage. Transmission direct, e.g. hookworms, *Haemonchus* (Nematoda), or indirect via an intermediate host, e.g. *Schistosoma* (Platyhelminthes).

Table 1.2 *Influence of factors involved in natural resistance and acquired resistance to parasitic infection*

Stage in host parasitic infection	Host characteristics that influence interaction			
	Behaviour	Structure	Physiology	Immunity
Initial contact	+	+	–	–
Establishment in host	–	+	+	–
Development of parasite	–	+	+	+
Reproduction of parasite	–	–	+	+

Notes:
Natural resistance is expressed through the host's behaviour, structure and physiology; acquired resistance is expressed through the host's immune response. The two categories are not absolute, each is intimately related to the other.

1.4   Natural resistance to infection

Natural resistance is a phenomenon operating at many levels (Table 1.2). Hosts can be considered naturally resistant to parasites because, through geographical distribution, behavioural characteristics or nutritional habits, they simply do not come into contact with infective stages. (Our natural resistance to may parasites is clearly of this kind and all too frequently breaks down.) Natural resistance more conventionally encompasses a physiological incompatibility between parasite and host environment that prevents invasion, establishment or survival without the intervention of immunologically based protective responses. For example:

(a) Helminths, such as schistosomes and hookworms, that rely on skin-penetrating larvae for infection may be unable successfully to cross the epidermis and basement membrane in order to enter the dermis.
(b) Trypanosomes entering the blood may be killed by factors naturally present in serum, such as the high density lipoprotein that destroys animal-infective *T.b. brucei* in humans, or may activate complement and be lysed.
(c) Intracellular protozoa such as *Plasmodium* may be unable to enter host cells because these lack essential surface molecules.
(d) Parasites that enter the host orally and which require activation by specific triggers in the intestinal environment may not experience these factors or the correct combination of factors that is necessary.

Even if the host provides an environment suitable for initial development, incompatibilities may well arise at subsequent stages through nutritional inadequacies, or through incomplete stimuli for migration and reproductive maturation. The eggs of the human *Ascaris, A. lumbricoides,* for example, can hatch in a variety of mammalian hosts because of the relatively low specificity of the intestinal signals required (mammalian body temperature, alkaline pH, dissolved CO_2 and a reducing environment). After hatching, the larvae can migrate and reach the lungs but development after this point, and sexual maturation, occur only in humans. It follows from this that, in natural host–parasite relationships, the parasite must be precisely adapted to the structural and physiological conditions that characterize the host species. This adaptation, which develops over long periods of evolutionary change, is the basis for the phenomenon of host specificity, i.e. the restriction of parasites to particular speces of hosts. In some cases the restriction is near absolute, as occurs with the human malaria parasites and certain of the human filarial nematodes. In other cases the restriction is extremely loose and parasites can undergo development in, and be transmitted between, a wide variety of hosts, as occurs with the nematode *Trichinella spiralis.* For the majority of parasites host specificity falls between these extremes and under natural conditions particular species are found in only a few species of hosts.

1.5 Acquired resistance to infection

In host–parasite relationships where natural resistance is low, the host can regulate the degree of infection to which it is subject only through the activities of its immune system, that is by developing an acquired resistance to infection. The degree to which such regulation operates, or indeed is necessary, under natural conditions is an open question and one for which few data are available. It is conceivable that in the wild, rates of transmission are such that parasite burdens remain below the thresholds necessary for the stimulation or expression of immune responses. With protozoan infections, of course, this factor is then offset by the parasite's ability to reproduce within the host. Rates of transmission are determined by a number of environmental factors, of which host density is one of the most important. It is the alterations of these factors, which humans impose upon themselves and their domestic animals, that lead to increased rates of transmission, increased prevalence and intensity of infection thus emphasizing the role of acquired immunity in control. Even where it is obvious that acquired immunity is necessary, the degree to which the host succeeds in regulating infection in this way is very variable. Many parasites have evolved successful means of evading the immune response that they evoke and some seem not to evoke protective responses at all. Others depend for their survival as a species on development and reproduction within individual hosts that are, for various reasons, incapable of developing or expressing an adequate protective immunity.

Implicit in the concepts of host specificity and of balanced relationships is the assumption that the ability of the host to control the parasite is never absolute. Several workers have put forward the concept that in evolution there is always a tendency towards a mutual adjustment between the two species, the parasite reducing its pathogenicity and immunogenicity so as to elicit weaker host responses, the host reducing its responsiveness so that parasite control is achieved without concomitant pathological change. 'Successful' parasites therefore tend to become harmless to the host. An example frequently cited in support of this concept is the relatively benign association of trypanosomes in game animals compared with the pathological consequences of human infections. Whether this interpretation is of general applicability is open to question. Many associations are still remarkably pathogenic for the host, in others the host rapidly achieves control of the parasite. Although it seems self-evident that mutual adjustment is necessary for mutual survival there are many parasites, particularly those transmitted through food chains, for which death or incapacity of the host can only facilitate their own transmission. It is, in fact, likely that there is no single trend in the evolution of host–parasite relations and that a variety of endpoints may be achieved, however three factors in particular suggest that the host–parasite relationship is always a confrontation for survival. One is the variability within species of the mechanisms controlling natural and acquired resistance. This implies the operation of strong selective pressures from infectious organisms. The second is the very existence and complexity of the resistance mechanisms possessed by all animals, but particularly by vertebrates. It is hard to see why such elaborate devices should have evolved, and why they should have persisted, if they did not confer powerful selective advantages. The third is the evidence of substantial genetic variability within parasite populations, which attests to the effectiveness of selection pressures exercised by host resistance.

2

The immune response

Protection against invading organisms

2.1 Introduction

The ability to discriminate between 'self' and 'not-self' is present throughout
the animal kingdom and indeed may be seen as necessary for the evolution of
complex animal organization as we know it. Recognition of 'not-self' certainly occurs in the Protozoa, where it is used in the selection of food, of com-
patible mating types and, in parasitic species, of appropriate host cells. The
importance of this ability, and its relevance for survival, increased enormously
with the evolution of multicellularity and the associated increases in size and
complexity. In multicellular organisms this ability has at least three major roles
to play in preserving the integrity of the body:

(*a*) in minimizing the consequences of contact with related but foreign (allo-
     geneic) organisms;
(*b*) in preventing the proliferation of mutant cells;
(*c*) in defence against invading pathogens.

The first of these is of significance only in primitive sessile animals, such as
sponges and corals, in which contact, through growth, with other species is a
regular occurrence. It is well known that these simple organisms possess efficient mechanisms of allogeneic recognition, capable of inducing destructive
cytotoxic reactions and thus preventing intermixing of colonies. It is remarkable that these recognition mechanisms also show the property of short-term
memory and enhanced effectiveness on re-presentation of the foreign stimulus. In higher animals recognition processes are primarily concerned with
defence against mutant cells and against pathogens. The importance of these

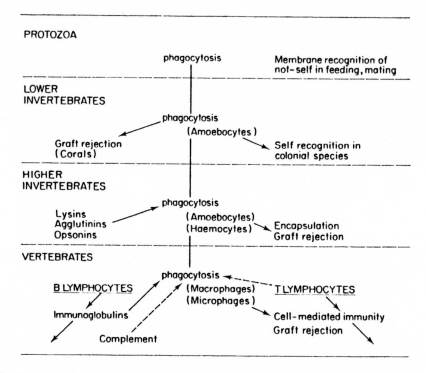

Fig. 2.1   Development of increasingly complex mechanisms of self – not-self discrimination in animal evolution.

processes, and the complexity of the mechanisms concerned with their operation, increases with the cellular complexity of the animal, reaching a peak in the warm-blooded vertebrates, i.e. in the birds and mammals (Fig. 2.1).

In lower invertebrates recognition of not-self is limited mainly to the detection of foreignness in molecules presented at the surfaces of cells. Expression of effector mechanisms is also primarily a surface phenomenon, mediated primarily by phagocytic cells although humoral factors may also be important. In vertebrates this basic ability is further elaborated, but is supplemented and overshadowed by the development of *adaptive immunity*, which allows a much greater sophistication and effectiveness in response. The improved capability stems from the evolution of a class of cells, the lymphocytes, whose homologies in invertebrates are controversial. Among the many important advantages conferred by the adaptive immune response are the ability:

(*a*) To discriminate not only between self and not-self, but also among an enormous variety of not-self, foreign molecules (*antigens*). [Although it is conventional to refer to foreign molecules as 'antigens', what is recognized as foreign is not the whole molecule but particular antigenic determinants or *epitopes* (discrete structural configurations) on that molecule. Complex molecules therefore present several epitopes each of which may be recognized by the immune system.]

(*b*) To retain long-term memory of foreignness and to respond more rapidly on subsequent contact (the *anamnestic response*).

(*c*) To produce large amounts of free receptor material (*antibody*) capable of combining with soluble or membrane-bound antigens.

(*d*) To interact with a number of non-specific resistance mechanisms, including the ancestral phagocytic cells.

(*e*) To confer protection against infection upon offspring.

Phagocytic cells still play important roles in resistance against infection, through their participation in inflammatory reactions, but their activities in warm-blooded hosts are so inextricably linked with immune responses that it is difficult to conceive of active anti-parasite responses that do not have a major immunological component.

2.2   The major histocompatibility complex

Implicit in the ability to recognize not-self is the recognition of self. This is an active process, i.e. it depends upon the presence of self molecules rather than the mere absence of not-self. A great deal is now known about the structure, production and control of such self markers. They were first identified in experiments involving grafting of tissues between related and unrelated animals, and for this reason they were initially described as *transplantation* or *histocompatibility* antigens. Involvement in graft recognition and rejection clearly represents an unnatural situation and there has been much speculation about the natural functions of these molecules. It is now known that they play key roles in the immune response, but their original designations have been retained.

The results of the early transplantation experiments showed that there were both major and minor histocompatibility antigens, i.e. antigens evoking strong and weak rejection responses when tissues were exchanged betweeen unrelated individuals. The gene loci coding for these antigens have been identified in many species, but are most completely known in mice and humans. In every species studied it has been found that the genes that code for the major histocompatibility antigens are grouped together on one particular chromo-

MOUSE [Based on BALB/c]

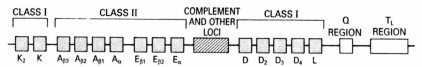

HUMAN

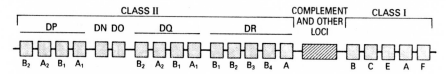

Fig. 2.2 Major histocompatibility complexes (MHC) of mice and humans. Many more genes have been identified in the MHC than are expressed. The figure shows some of the loci that are most important in immune responses. Class I gene products, the classical transplantation antigens, are expressed on almost all cells. Expression of Class II products is much more restricted (e.g. macrophages, dendritic cells, B cells). (Based on Klein, 1990, *Immunology*, Blackwell Scientific Publications).

some, forming the major histocompatibility complex (MHC). The genes within the MHC are grouped closely together and so appear to segregate in a simple Mendelian fashion. Recombinations between the genes do occur, however, and can be produced readily using defined strains of mice.

In humans the MHC is located on chromosome 6 and is known as the HLA (human leucocyte antigen) complex. In mice the MHC is located on chromosome 17 and is known as the H-2 (histocompatibility antigen-2) complex. The gene loci in the MHC are grouped into three classes (Fig. 2.2). All are concerned with immune functions, but two (Class I and Class II) play major roles in interactions involving lymphocytes and antigens. Genes in both classes code for molecules that are expressed as integral components of cell membranes, but there are class differences in the cell populations concerned and in the structure of the gene products. Molecules coded by Class I loci occur on virtually all cells, whereas Class II-coded molecules have a more restricted distribution, occurring on cells such as macrophages, dendritic cells, Langerhans cells, B lymphocytes and some T lymphocytes. The consequences of these class-restricted patterns of distribution will be discussed later.

The Class I and Class II gene products are heterodimers, i.e. they have a two chain structure, the two chains being dissimilar. In Class I molecules the chains are very different. One, the β microglobulin chain, is not MHC-coded, but is necessary for the correct functioning of the other, the MHC-coded α chain.

This has three domains. The terminal domains are made up of variable sequences of amino acids and create a cup into which antigen fragments can bind. Class II molecules are built up of similar α and β chains, both coded by MHC loci. Each chain consists of variable and constant domains, the former again creating a site in which antigen fragments can bind.

2.3 Lymphocytes

The basic patterns of lymphocyte development and differentiation are summarized in Fig. 2.3. In development, lymphocyte precursors arise from yolk sac cells and foetal liver, but subsequently all lymphocytes develop from stem cells that originate in the bone marrow. Two major pathways of differentiation are then followed, which lead to the production of two classes of lymphocytes, one concerned primarily with helper function and cell-mediated immune responses, the other with antibody-mediated responses.

In the first pathway, precursor cells pass to the thymus where they proliferate and mature to become thymus-dependent or T lymphocytes. The process of maturation involves the expression of receptors capable of binding with antigen, the selection of cells capable of recognizing self MHC and foreign antigens, and the deletion of those recognizing self antigens. During this process cells acquire cell surface molecules that help to determine their future roles in the immune response and so define a number of T cell subsets. The major subsets are those expressing the CD4 marker, which function primarily as helper cells, and those carrying the CD8 marker, which function as cytotoxic cells. (CD stands for 'cluster of differentiation' and is used to describe molecules that can be used to identify different populations of leucocytes. A very large number of CD molecules is now known.) Mature T cells leave the thymus and pass to the secondary lymphoid organs, the spleen and lymph nodes, forming part of the lymphocyte population located in these organs or recirculating around the body. The second pathway of lymphocyte differentiation is seen most clearly in birds, where it involves a defined primary lymphoid structure, the bursa of Fabricius, not present in mammals. Lymphocytes differentiating in this organ are referred to as B (bursa dependent) cells and like T cells, become distributed among the secondary lymphoid organs. In mammals differentiation of B cells takes place in the bone marrow.

As they differentiate, T and B cells acquire a variety of surface molecules that serve as characteristic markers and that play important roles in their cellular functions. These markers coexist with molecules coded for by MHC genes. Both T and B cells carry membrane receptors capable of combining specif-

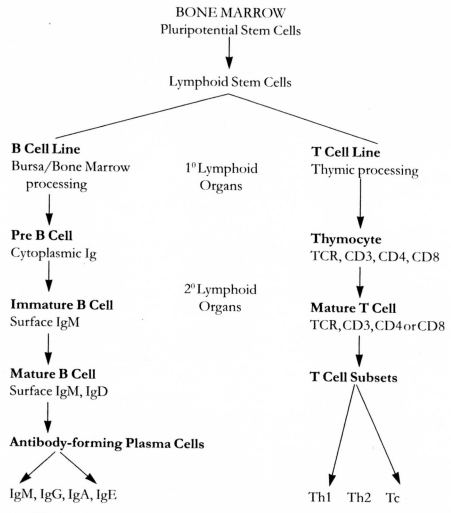

Fig. 2.3   Development and differentiation of lymphocytes. Th, T helper; Tc, T cytotoxic.

ically with antigen, the progeny of each clone of cells carrying receptors for one particular epitope. The nature of the receptor differs between the two classes of lymphocyte. In B cells the receptor is immunoglobulin of the IgM class, the T cell receptor (TCR) is a heterodimeric molecule that forms a complex (the CD3 complex) with a number of other molecules. In most T cells the TCR is made up of an alpha and a beta chain, but in certain subsets of T cells the TCR has gamma-delta chains.

T and B cells also differ in some of their membrane glycoproteins, and therefore combine differently with particular lectins (molecules of plant origin that bind to specific sugars). Exposure of lymphocytes to lectins triggers transformation and cell division. Measurement of responses to mitogens, together with identification of surface receptors, provide convenient means of identifying and differentiating between the two lymphocyte populations.

The secondary lymphoid organs have a characteristic structural organization (Fig. 2.4) within which T and B cells are distributed in precisely defined regions. Lymph and lymphocytes flow through these organs, draining specific parts of the body and returning to the general circulation via the thoracic duct. Lymphocytes also enter the spleen and lymph nodes from the blood, passing across the endothelia of special post-capillary vessels, the high endothelium venules (HEV). This traffic is made possible because lymphocytes express adhesion molecules, surface receptors specific for ligands present on the surface of the HEV cells. Having entered, the lymphocytes move through the lymphoid organ and eventually rejoin the efferent lymph.

2.4   Recognition of antigen and initiation of the immune response

2.4.1   CD4+ T helper cells

Initiation of an immune response almost always requires that antigens are processed and presented to T helper cells by accessory cells; direct contact between antigens and lymphocytes can in fact induce unresponsiveness. Processing and presentation can be achieved by a variety of cells expressing MHC Class II, but it is likely that macrophages, dendritic cells, Langerhans cells and B cells play the major role. Antigens are taken up by these cells and processed, i.e. undergo enzymatic breakdown, in endosomes (Fig. 2.5). Small fragments of the antigen (for peptides, usually sequences of about 10 amino acids) then bind to Class II MHC molecules present within the endosome, and are protected from further breakdown. These bound fragments (antigenic determinants or epitopes) are eventually re-expressed at the cell surface as an MHC-epitope complex and in this form the epitope can be recognized by the TCR. Binding of the epitope to the TCR provides an initial signal (signal 1) to the T cell, and other molecular interactions, e.g. between CD28 and B7 (Fig. 2.6), provide the co-stimulatory signal 2 which allows the T cell to respond. Antigen recognition is highly specific; TCR can discriminate between epitopes that differ in only one amino acid.

Once recognition has occurred the T cell concerned transforms into a blast

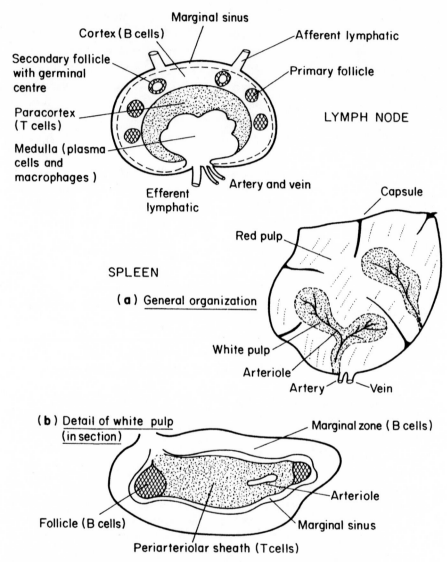

Fig. 2.4   Structures of lymph nodes and spleen, showing distribution of T and B lymphocyte areas.

cell, becoming larger, metabolically more active and entering the cell cycle. This response is triggered by the molecular interactions between the TCR and the epitope, and between the accessory molecules on the APC and T cell, but is also dependent upon soluble factors (cytokines) released initially from the APC. The responding T cell undergoes repeated divisions to produce a clone

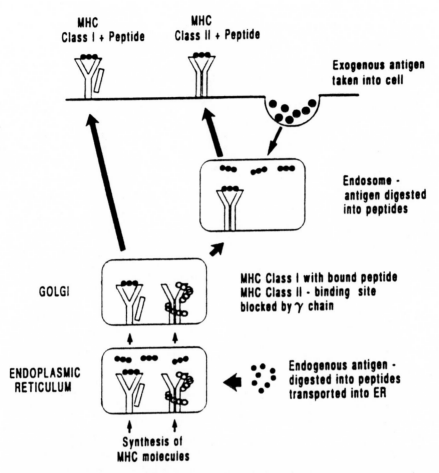

Fig. 2.5 Processing of exogenous and endogenous antigens in antigen presenting cells. Exogenous peptides bind to Class II MHC molecules in the endosomes, endogenous peptides bind to Class I MHC molecules in the endoplasmic reticulum. Binding of peptides to Class II molecules is blocked until the gamma chain is removed. (Taken from Smyth, 1994, *Introduction to Animal Parasitology*, Cambridge University Press.)

of cells, each carrying the TCR specific for the initial epitope. The fate and function of these cells varies considerably. Some revert to the small resting phase as memory cells, providing the cellular basis for enhanced secondary responsiveness. Others continue to function as helpers, releasing a variety of cytokines, and interacting with other T cells, with B cells and with myeloid cells to generate the effector phases of the immune response.

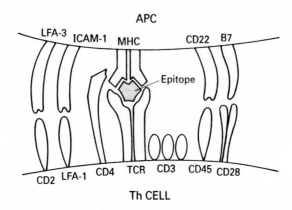

Fig. 2.6   Schematic representation of molecular interactions between antigen presenting cell (APC) and T helper cell (Th) during antigen recognition. The epitope is presented by the Class II MHC molecule and recognized by the T cell receptor (TCR). Signal transduction involves the CD3 and CD4 molecules. A number of additional ligand–receptor interactions facilitate the APC–Th cell interaction.

2.4.2   CD8⁺ T cytotoxic cells

MHC Class II molecules are essential for presentation of antigens to T helper cells. Only certain cells of the body express Class II, and only these cells can function as antigen presenting cells for CD4⁺ T cells. In contrast, presentation to CD8⁺ T cells requires MHC Class I molecules. These are expressed on almost all body cells, and therefore almost all cells can present antigen in this way. Presentation on the two classes of MHC molecules reflects two separate routes of processing in the cell concerned (Fig. 2.5). Antigens entering cells from the outside (exogenous antigens – e.g. from extracellular parasites) are processed and combine with Class II molecules in endosomes; antigens arising within the cell itself (endogenous antigens – e.g. from intracellular parasites) are processed cytoplasmically and combine with Class I molecules in the endoplasmic reticulum, eventually appearing at the cell surface. With help from CD4⁺ cells, CD8⁺ T cells responding to such antigens can mature into cytotoxic killer cells, capable of combining with and lysing any cell expressing the same antigen–MHC combination. Cytotoxicity has not been considered as a major effector mechanism against parasites, but recent work has shown an important role against a number of intracellular protozoans, including the exoerythrocytic stages of malaria (see Chapter 4).

CD8⁺ T cells show MHC-restricted cytotoxicity. A population of lymphocyte-like cells – natural killer (NK) cells – are not antigen specific in their cyto-toxicity. NK cells have some of the characteristics of T cells. They participate in ADCC responses (see p. 26) but can also kill cells directly, this activity being regulated by stimulatory and inhibitory receptors. NK cells are known to play a part in defence against some parasites.

2.5   Cytokines

The development and activities of immune and inflammatory cells are regulated by interactions with a complex network of membrane-bound and soluble molecules collectively called cytokines (interleukins/lymphokines/monokines – Table 2.1). These are produced by a wide variety of cells, including lymphocytes and macrophages, and act on their targets through membrane receptors. Cytokine–receptor binding often leads to the production of more cytokine molecules and the expression of more receptors, thus producing a response cascade. Their action is often short range, but cytokines do circulate in the blood and can act at a distance. Target cells can be the producing cell (autocrine activity) or other cells (paracrine activity) and may be members of the same cell type or be quite different. For example, as T helper cells respond to antigen they are stimulated to release Interleukin 2 (IL-2) and at the same time to express receptors for IL-2. Binding of IL-2 to IL-2 receptors stimulates further T cell activity, the release of more cytokine and the expression of more receptor molecules – autocrine activity. Later on the T cells may release a wide variety of cytokines, which serve to regulate the developing response by acting upon other immune and inflammatory cells, either locally or remotely (e.g. the bone) – paracrine activity. In the mouse, and in humans, individual T helper (Th) cells tend to release one or other of a group of cytokines (type 1 or 2) and can be grouped into one of three subsets – Th0, Th1 and Th2 (Table 2.2). The bias of Th cells towards one or other of these subsets determines the eventual outcome of the immune response both qualitatively (in terms of effectors) and quantitatively. In very general terms Th1 cytokines promote the elements of cellular immunity, such as macrophage activation, and Th2 cytokines promote humoral responses.

2.6   Antibody production and immunoglobulins

Not all antibody production is dependent upon prior help from T cells. Certain molecules, the thymus-independent antigens found for example in

Table 2.1 *Major cytokines relevant to immunoparasitology*

Cytokine	Main Source	Functions
IL-1	Many cells	Activation, regulation, inflammation
IL-2	T cells	Stimulates T cells, B cells, macrophages, T cell proliferation
IL-3	T cells	Stimulates B cells; multi-potential CSF for many cell types in bone marrow
IL-4	T cells	Stimulates T and B cells; induces IgE; mast cell development; enhances MHC expression; down-regulates Th1 cells
IL-5	T cells	Stimulates B cells; induces IgA; eosinophil development
IL-6	Many cells	Stimulates T cells, B cells and granulocytes; induces acute phase proteins
IL-9	T cells	Mast cell development
IL-10	T cells	Down-regulates Th1 cells
IL-12	Macrophages	Stimulates release of IFN and B cells
IL-13	T cells	Suppresses macrophage cytotoxicity, down-regulates production of IFN-γ and IL-12
IFN–γ	T cells, NK cells	Activates macrophages; enhances MHC expression; stimulates B cells; induces acute phase proteins, down-regulates Th2 cells
TNF-α	Macrophages	Inflammation, cytotoxicity, cytokine release
TNF-β	T cells	Inflammation, cytotoxicity, cytokine release
CSF	Many cells	Control production of myeloid cell colonies in bone marrow. (e.g. GM-CSF, granulocyte /macrophage; G-CSF, granulocyte)

Notes:
IL, interleukin; IFN, interferon; TNF, tumour necrosis factor; CSF, colony stimulating factor.

some bacteria, can bypass this requirement and stimulate B cells directly. It is characteristic of such antigens, the majority of which are carbohydrate, that they possess serially repeated epitopes and therefore cross-link the immunoglobulin molecules that function as antigen receptors on the B cell surface. This cross-linking results in B cell activation. In essence, the role of T helper cells in inducing antibody responses to T-dependent antigens is also to provide a mechanism for cross-linking surface immunoglobulin. An older hypothesis was that T cells and B cells specific for the same antigen became linked by an antigen bridge, the TCR and immunoglobulin recognizing differ-

Table 2.2 *Major cytokines associated with T helper cell subsets (based on data from* in vivo *and* in vitro *studies with mice)*

	Type 1		Type 2					Type 3		
	IFN-γ	IL-2	IL-4	IL-5	IL-6	IL-10	IL-13	IL-3	GM-CSF	TNF-β
T helper 1	+	+	−	−	−	−	−	+	+	+
T helper 2	−	−	+	+	+	+	+	+	+	+
T helper 0	+	+	+	+	+	+	+	+	+	+

(Interleukins IL-1, IL-7, IL-8, IL-11, IL-12, IL-14, IL-15 are produced by non-T cells)

ent epitopes on the antigen molecule (carrier and hapten respectively). With the realization that B cells can process and present antigen, a different hypothesis is that the link is made because the B cell can present on its surface, complexed with MHC class II, the epitope for which the TCR is specific. Whatever the interaction, cell–cell contact is not sufficient by itself, production of antibody also requires co-stimulation by the appropriate T cell cytokines.

Following T cell triggering, the B cell enlarges, enters the division cycle, and becomes a B lymphoblast or plasmablast. Progeny of the dividing cells may remain large, differentiate into plasma cells and secrete immunoglobulin (Ig) or become small memory cells. Plasma cells initially release IgM, but then may switch to other Ig classes (isotypes), the switch being under the control of T cell cytokines. Each cell secretes Ig specific for one epitope, and this specificity remains despite the changes in isotype.

2.6.1   Immunoglobulins

Development of the ability to produce Igs can be seen as one of the major steps in the evolution of the adaptive immune response. It represents a facet of the vertebrate response that is essentially without prarallel in the invertebrates. An instructive way of thinking of these molecules is as cell surface receptors, produced in excessive quantities and released into the circulation. Here they can fulfill their recognition function at some distance from the source of production. This makes possible the localization of organs in which plasma cells can be concentrated, without losing the ability to provide the whole body with efficient protection.

Recognition and combination with antigens is the primary function of Igs, but combination is then followed by a variety of biological effects, the nature

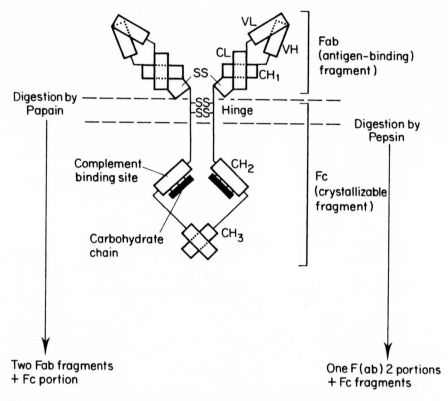

Fig. 2.7   Representation of an IgG$_1$ molecule, based on the Edelman and Porter model. The molecule is formed of two identical light chains (MW 25kDa) and two identical heavy chains (MW50 kDa) held together by disulphide bridges. The light chains are folded into two domains VL and CL, the heavy chains into four domains VH, CH$_1$, CH$_2$ and CH$_3$, and all except CH$_2$ are arranged in pairs held together by non-covalent forces. The amino acid sequences of the domains are variable (V) or constant (C) and the antigen binding sites are associated with hypervariable regions in the VL and VH domains. Digestion by proteolytic enzymes breaks the molecule into a variety of fragments. The Fab fragments contain the active antigen-binding sites, the Fc portion determines the biological characteristics of the molecule.

of which depends upon the isotype involved. In order to explain this point more fully it is necessary to understand in outline the structure of Ig molecules and to be aware of the differences that exist between isotypes.

The basic structure of Ig can be seen in the IgG molecule, which is composed of two identical light chains and two identical heavy chains, held together by a number of disulphide bonds (Fig. 2.7). Each chain consists of regions where amino acid sequences are constant between different molecules

Table 2.3 *Characteristics of immunoglobulins*

Isotype	Structure	MW (kDa)	% total serum Ig	Characteristics
IgM	Pentamer	900	6	First isotype in response, good agglutinator, fixes complement, secreted across mucosal surfaces
IgG	Monomer	150	80	Major isotype in body fluids, fixes complement, binds to macrophages and polymorphs, facilitates ADCC, crosses placenta,several subclasses
IgA	Monomer Dimer (+ secretory piece)	160 400	13	Major isotype at mucosal surfaces, secreted across epithelial cells, secreted in milk, binds to eosinophils, can activate complement by alternative pathway
IgE	Monomer	200	0.002	Binds to mast cells and basophils, involved in immediate hypersensitivity, binds to eosinophils, facilitates ADCC

of the Ig class (the constant regions) and regions where the sequences are variable. The variable regions lie at the N-terminal ends of the light and heavy chains and contain within them areas where there is hypervariability in sequences.

The Ig molecule can be split, by papain digestion, into three units, two Fab fragments, consisting of variable and constant regions of the light and heavy chains, and one Fc fragment, which consists of the constant regions of the heavy chains. Antigen specificity resides in the Fab fragments and is associated with the hypervariable regions. These form structural configurations that are complementary to those of the appropriate antigenic determinants. The Fc fragment (so called because it can by crystallized) determines the biological properties of the Ig molecule and these are characteristic for each isotype (Table 2.3). Activation of complement, which is discussed further below, depends upon a specific region (domain) of the Fc fragment, the configuration of which alters after antigen–antibody combination. Activation is most efficient with IgM, because the pentamerous structure of the molecule results in fixation of a proportionately greater number of complement molecules than occurs with monomeric Igs.

With the exception of IgD, all Igs can readily be identified in blood and tissue fluids. IgM and IgA can also be actively secreted across mucosal membranes and thus form the major Igs in the normal intestinal lumen. Secretion occurs across the epithelial cells of the mucosae and, in the case of IgA, across cells lining bile ducts, and requires the addition to the Ig molecule of a secretory piece. This both facilitates trans-epithelial passage and protects intramolecular linkages from the action of enzymes. In this way the functional life of the Ig is prolonged in the proteolytic environment of the intestine.

2.6.2   Immunoglobulin function

The interaction of Ig with parasite antigens can have many different consequences. Binding to soluble antigens can interfere with their biological functions, e.g. toxins are neutralized and enzymes are inactivated. Cross-linkage of antigens present on protozoa may cause them to agglutinate, rendering them non-infective. Coating of parasite surfaces by antibodies – the process of *opsonization* – facilitates phagocytosis if the target is of suitable size, the phagocytic cell binding via its membrane Fc receptors. If phagocytosis is impossible, because the target is too large, cells bind and adhere closely to the surface. Adherence may then be followed by release of lysosomal enzymes, granule contents and other cytotoxic factors, resulting in surface damage – the process known as antibody-dependent cell-mediated cytotoxicity (ADCC). Many cell types (macrophages, neutrophils, eosinophils, platelets, NK cells) can participate in ADCC reactions (Fig. 2.8), their adherence to the antibody-coated parasite again being mediated through Fc receptors (mainly for IgG and IgE, although it is now known that IgA can also be involved). Cells can also be armed for ADCC by binding of immune complexes to their Fc receptors, free antibody-combining sites in the complex allowing specific recognition of target antigens.

The interaction of antigen with IgM or IgG molecules makes available on the Fc region a site to which the initial components of *complement* can attach. Complement is the name given to a complex of proteins present in serum, capable of being activated to become enzymatically active. When complement becomes attached (fixed) to an antigen–antibody complex a sequence of reactions is inititated, each component acting enzymatically to activate subsequent components in a cascade fashion. Activation of complement can also occur in the absence of immune complexes, the cascade being triggered, for example, by molecules present on the surfaces of infectious organisms including parasites. The stages involved in the classical (complex-initiated) and alternative (pathogen-initiated) pathways of complement activation are shown in Fig. 2.9.

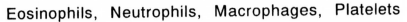

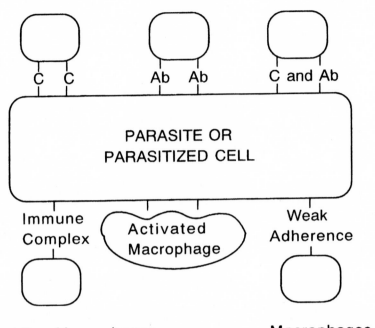

Fig. 2.8 Summary diagram showing the variety of interactions that are possible between the surfaces of parasites or parasitized cells and myeloid cells (eosinophils, macrophages, neutrophils, platelets). Cells may bind to the target by their complement (C) or antibody (Ab – IgM, IgG, IgA, IgE) receptors after complement activation and/or antibody binding; bound immune complexes allow antigen-specific binding. Binding also occurs by intermolecular interactions (e.g. sugars–lectins) the strength of which is increased when cells are activated. Interactions may result in phagocytosis, if the target is small, or ADCC if it is large.

Many of the activation products of complement are biologically active. They mediate chemotaxis of inflammatory cells and cell adherence, they cause release of histamine and bring about vasodilation, and the end product – the membrane attack complex – can cause lysis of target cells after insertion into the plasma membrane. Some of the most important products in the context of immunoparasitology are C3b and its inactive derivative iC3b, which attach to parasite membranes allowing adherence of cells bearing C3b/iC3b receptors (e.g. CR1 for C3b on macrophages and eosinophils, CR3 and CR4 for iC3b on macrophages), and C3a and C5a, which bind to cells carrying the

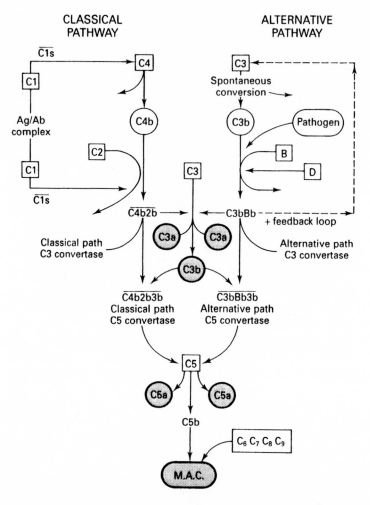

Fig. 2.9 Simplified scheme of complement activation by the classical and alternative pathways. □, complement component; ○, cleavage product; ●, biologically active product; C1s etc., enzymatically active form; M.A.C., membrane attack complex.

appropriate receptors and promote chemotactic and anaphylatoxic (histamine-releasing) activity.

In addition to its involvement in ADCC through low affinity Fc receptors on cells such as macrophages, eosinophils and platelets, IgE antibody plays a major role in inflammatory responses mediated by the release of potent biologically active mediators from mast cells and basophils. These carry high-affinity receptors for IgE and bind it avidly. When cells that have bound antigen-specific IgE are subsequently exposed to the antigen concerned, the antibody molecules are cross-linked through their Fab regions and the cells degranulate. The mediators released then initiate the tissue inflammatory reactions characteristic of the immediate hypersensitivity response, best known in allergic conditions such as hay fever or asthma.

2.7   Interactions between lymphocytes and myeloid cells

The vertebrate immune response to infection is an integrated response, involving not only the cells of the immune system proper, i.e. the lymphocytes, but also a variety of other cells, the myeloid cells, which, like lymphocytes, are derived from stem cells originating in the bone marrow. Three distinct groups of myeloid cells can be considered, the macrophage–monocyte series, the granulocytes and the amine-containing cells.

2.7.1   Macrophage–monocyte series

Monocytes circulate in the blood and are the precursors of macrophages. Macrophages represent a direct link with the scavenging phagocytic cells of more primitive animals and are still important in this respect. However, in vertebrates they are also involved in the induction, regulation and expression of immune responses. This activity reflects their capacity to process and present antigen, to release a variety of cytokines and to interact with the surfaces of malignant cells or parasites. Many aspects of macrophage function depend on the presence of cell membrane molecules, including MHC gene products and a variety of receptors, more than 50 of which have now been defined, including the Fc and C3b receptors involved in phagocytosis.

Macrophage activity is enhanced by cytokines released from T cells (e.g. IFN-γ) and by factors present in invading pathogens (e.g. LPS – lipopolysaccharide). Activated cells respond by becoming more motile, phagocytose or adhere more readily and digest more efficiently when material is taken up into the phagolysosomes. They also release the pleiotropic (multi-function) cytokine IL-1 and express more MHC Class II molecules at the cell surface. In this activated state, macrophage function as effector cells may be expressed non-specifically, i.e. although activation follows a specific stimulus, such as an immune response to organism A, the cells will then readily attack any other organisms (B, C, D, etc) accessible to them. Specificity in macrophage activity can be achieved by arming with antibody or immune complexes. The armed

Table 2.4 *Factors released from macrophages that interact with components of the immune and inflammatory systems*

Coagulation factors
Complement components
Cytokines (e.g. IL-1, IFN-γ, TNF-α)
Enzymes (acid and neutral hydrolases, lysozyme)
Enzyme inhibitors
Fibronectin
Inflammatory mediators (leukotrienes, prostaglandins, PAF)
Reactive metabolites (nitric oxide, oxygen radicals)

and activated cell is then able to express a number of specific effector functions, including ADCC. The ability of these cells to damage or kill parasites during contact, or after phagocytosis, depends on their production of a variety of factors (Table 2.4), of which lysosomal enzymes, oxygen metabolites, nitric oxide and tumour necrosis factor (TNF) are among the most important. The production of oxygen metabolites such as singlet oxygen ($^1O_2$), hydrogen peroxide (H_2O_2) and superoxide (O_2^-) follows the burst of oxygen consumption – the oxidative burst – linked to the process of phagocytosis, which is much more efficient in activated cells.

2.7.2 Granulocytes

Eosinophils and neutrophils, present both in blood and in body tissue, are characterized by polymorphic nuclei and prominent cytoplasmic granules. Neutrophils are sometimes referred to as microphages and have as a primary function the phagocytosis and destruction of microorganisms. They possess both Fc and C3b receptors and, as a result, their efficiency in phagocytosis is enhanced when specific antibody is bound to the organism or if the surface of the organism itself activates complement. After uptake, the organism is killed in the phagolysosome by lysozyme, lysosomal enzymes, cationic proteins and oxygen metabolites, and some of these factors may also be released by exocytosis into the extracellular environment. Neutrophils are positively attracted by C3a, one of the products of complement activation, as well as by lymphocyte and mast cell factors. This chemotaxis plays an important role in directing defensive responses. Unfortunately this attraction also occurs wherever antigen–antibody complexes form, and the accumulation of neutrophils, together with the extracellular release of their enzymes, contributes to tissue inflammation.

The role played by eosinophils has been obscure, although their association with the inflammatory responses and parasitic infections has been known for many years. Clarification of their function has come largely from recent studies in their *in vitro* interactions with helminths. The characteristic granules of these cells contain a variety of hydrolytic enzymes, including peroxidases, and prominent central bodies that contain cationic proteins. It has been shown that the eosinophil membrane bears both Fc and C3b receptors, which make it possible for the cell to adhere to target surfaces coated with antibody or complement components. An important finding is that in addition to Fc receptors for IgG, low affinity Fc receptors (FcεRII) exist for IgE. Adherence is followed by release from the granules of enzymes, the major basic protein of the granule core, and other factors that readily damage cell membranes. The production and activities of eosinophils are regulated by T lymphocytes through cytokines, especially IL-5 and, additionally, by complex interactions with amine-containing cells. The latter release a powerful eosinophil chemotactic factor (ECF-A), which attracts eosinophils into sites of amine release and may increase expression of their membrane receptors, thus enhancing their activities. Eosinophils in turn release factors that counteract the activity of released amines. A recent and significant observation is that eosinophils can produce a variety of cytokines.

2.7.3   Amine-containing cells

These include the mast cells and basophils, two cell types with many functional similarities, but with some basic differences. Basophils circulate in the blood and occur in the tissues; mast cells occur only in the tissues and form two distinct populations, one restricted to mucosal surfaces, the other distributed in connective tissues. The myeloid origin of basophils and their relation to granulocytes have never been in question; the demonstration that mast cells are similarly derived is more recent. Both cells are heavily granulated and contain a variety of potent mediators, including histamine, 5-hydroxytryptamine (serotonin), heparin, enzymes and chemotactic factors. The membranes of both cells carry high-affinity receptors for the Fc region of IgE (FcεR1) and particular IgG subclasses. Crosslinking of membrane-bound Ig by antigen triggers exocytosis of the granules and release of their contents. Amine-containing cells are typically associated with immediate hypersensitivity reactions (see below) but they also play a role in other immune events. Like

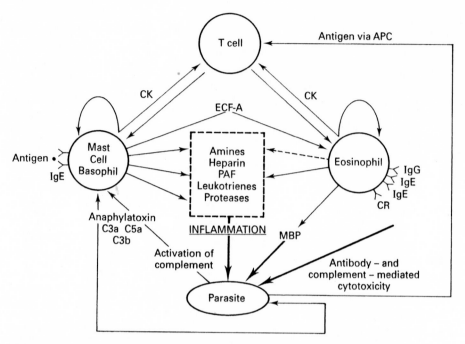

Fig. 2.10   Interactions between lymphocytes, amine-containing cells, eosinophils and parasites. →, factors released by cells; --→, inactivating factors released by eosinophils; →, anti-parasite responses; APC, antigen presenting cell; C3 etc., complement components; CK, cytokines; CR, complement receptor; ECF-A, eosinophil chemotactic factor of anaphylaxis; Ig, immunoglobulin; PAF, platelet activating factor.

eosinophils, basophils and mast cells are known to produce a number of cytokines and may play an important role in polarizing T helper cell responses.

The interactions between lymphocytes, amine-containing cells, eosinophils and parasites are summarized in Fig. 2.10.

2.8   Hypersensitivity

Under certain conditions, the normal secondary response to contact with a previously experienced antigen is associated with exaggerated reactions that may result in tissue damage. Some are rapid, explosive responses, others are delayed in development, all are referred to as hypersensitivity reactions. Such reactions were originally categorized by Coombs and Gell into Types I, II, III and IV (Fig. 2.11). The first three are sometimes

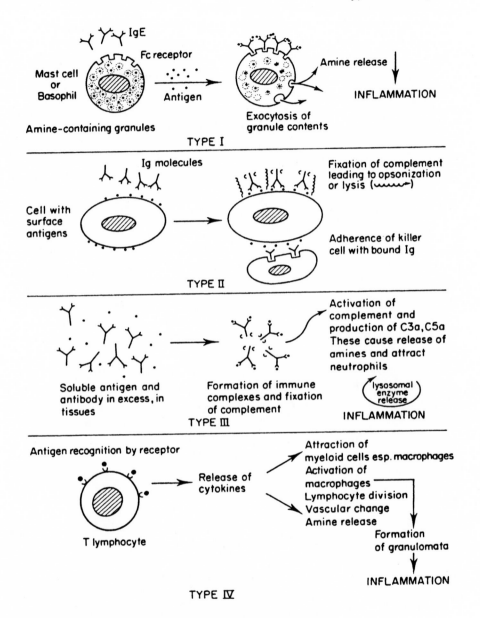

Fig. 2.11   Summary of hypersensitivity reactions.
Type I: Immediate-type hypersensitivity, initiated by IgE antibody.
Type II: Antibody-dependent cytotoxic hypersensitivity, resulting in cell lysis.
Type III: Complex-mediated hypersensitivity, resulting in the Arthus reaction in tissues, or in serum sickness.
Type IV: Delayed-type hypersensitivity: resulting in granuloma formation.

collectivley referred to as the immediate hypersensitivity reactions because of the speed of onset, but this term is commonly restricted to Type I, which is dependent upon interactions between antibodies and amine-containing cells. Type II reactions are antibody-dependent cytotoxic phenomena (ADCC) and Type III are antigen–antibody complex-mediated reactions. Type IV are delayed-type hypersensitivity reactions (DTH) that are distinctive not only in their slower time course, but also in their primary reliance upon T cells rather than upon antibody. Types I, III and IV are considered briefly below.

2.8.1   Immediate hypersensitivity

Reactions are triggered when antigens cross-link antibody molecules bound to receptors on the surface of mast cells or basophils. Their effects are entirely due to the release of potent mediators that bring about immediate physiological changes such as increased contractility of smooth muscle and increased permeability of blood vessels and epithelial membranes. In severe cases the reaction is a generalized anaphylaxis with systemic involvement. More often the reaction is localized, affecting parts of the body that meet environmental antigens, e.g. skin and mucous membranes. An important consequence of mediator release and increased permeability is the escape of plasma and cells into inflamed tissues or into the lumen of organs such as the intestine. The reactions are modulated by factors released from other cells, notably eosinophils, which rapidly inactivate the mediators responsible. Eosinophils are in fact selectively recruited to sites where Type I reactions are in progress through the release of chemotactic factors.

2.8.2   Immune complex-mediated hypersensitivity

Fixation of complement by antigen–antibody complexes results in the liberation of components which, in combination, lead to the generation of inflammatory responses. C3a and C5a cause release of histamine, thus increasing vascular permeability locally, they also attract neutrophils, which phagocytose complexes and in so doing release a variety of factors, including enzymes, that damage surrounding tissues. If uncontrolled, inflammation intensifies in a vicious circle. When immune complexes are formed in antibody excess they are readily precipitated and complex-mediated damage is localized – the Arthus type reaction. In antigen excess, complexes remain soluble and circulate around the body before being deposited in various tissues, which then become the focus of inflammatory lesions.

2.8.3   Delayed-type hypersensitivity

DTH is a correlate of the cell-mediated immunity that protects against a number of intracellular infections, notably those of bacterial origin. In its classical form, e.g. in the response to tuberculosis, it is initiated by a subset of T cells (Th1 Cells) which, after recognition of antigen, release cytokines such as IFN-γ that attract, localize and activate macrophages at the site of antigenic stimulation. Characteristically the inflammatory site becomes populated by mononuclear cells, i.e. lymphocytes and macrophages, rather than polymorphonuclear cells. Neutrophils appear only in the initial stages, eosinophils may appear later. The development of cytotoxicity in T cells, as is seen during graft rejection, can also be considered a form of DTH, and is closely linked to the appearance of armed macrophages, which are also cytotoxic in this situation. A specialized form of DTH is the Jones–Mote reaction, or cutaneous basophil hypersensitivity, in which, as the name implies, inflammatory sites in the skin are infiltrated by basophils. This reaction appears to be an important defensive response against ectoparasitic arthropods (Chapter 9).

Although each type of hypersensitivity reaction has been considered separately, it is likely that there is a complex interplay between them in any hypersensitivity response. Certainly Type I reactions are now thought to form an important component in the development of Type III and Type IV, primarily by facilitating extravasation and accumulation of effector cells at sites of injury and antigen deposition. This aspect is clearly relevant to anti-parasite responses and will be discussed in greater detail in later chapters.

3

Experimental immunoparasitology

3.1 Introduction

Immunity to parasites, in the sense of protection from infection, or from the consequences of infection, has been recognized empirically for a very long time, even though the underlying causes have not been understood. Anti-parasite immunity was originally approached very much in the same terms as classical anti-microbial immunity. However, it is now clear that many significant differences exist, in part related to the greater size and complexity of parasites. Immunoparasitology as a distinct discipline is a comparatively recent development and it is only in the last 20 years or so that substantial progress has been made. One of the most important factors in encouraging work in this field has been the World Bank/WHO programme of research into parasitic diseases of man. This has stimulated a much greater awareness of the problems caused by parasites and has encouraged cooperation between parasitologists and immunologists in searching for improved methods of immunologically based control. An essential part of this cooperation has been the investigation of fundamental aspects of the immunological relationship between hosts and parasites.

Parasites present a number of difficulties for the experimentalist concerned with analysis of immunity. These can be considered under three headings:

(*a*) parasite maintenance in the laboratory;
(*b*) antigenic complexity;
(*c*) identification and measurement of immunity.

3.2   Parasite maintenance

Immunological analysis of *in vivo* responses are most conveniently carried out using well-known experimental animals, in which variables can be fully controlled. Many parasites of medical or veterinary importance (e.g. human malarias, filarial nematodes) show rigid host specificity and cannot readily be passaged in laboratory hosts. The choice is then between studying the parasite in its natural host, with the attendant difficulties in experimentation, or using related parasites that can be maintained in the laboratory, but which may not always be entirely appropriate as model systems. A third possibility is to use *in vitro* maintenance. Some significant progress has been made in this field and the availability of techniques for long-term culture of, for example, *Plasmodium falciparum* and bloodstream form trypanosomes, has vastly increased research potential. However, such techniques are available for relatively few species, the restrictions being particularly acute with helminths. Some, e.g. schistosomes, larval nematodes, can be kept successfully in short-term cultures, but few can be maintained for extended periods or taken through their complete life cycles. Of equal significance is the problem of modelling *in vitro* the conditions associated with the expression of immunity in the host. For some parasites, e.g. gastro-intestinal species, where immunity is expressed primarily in expulsive mechanisms, it is unlikely that this will ever be achieved.

3.3   Parasite antigens

Parasites are complex organisms and often have complex developmental cycles. As a result they present both host and immunoparasitologist with a battery of antigenic material, much of which is undefined. There are, therefore, difficulties in relating particular components of a parasite to the generation of immune responses, and of differentiating between antigens that elicit protective reponses and antigens that have no apparent role in resistance, i.e. between 'functional' and 'non-functional' antigens. A useful, though arbitrary, distinction has been made between antigens that are associated with structural components of a parasite (structural antigens) and those that are released as a result of surface turnover or of metabolic processes (excretory–secretory [ES] antigens). Many attempts have been made to immunize animals against infection using homogenates of parasite material. Although some have been successful, the majority have not, even though infection with the parasite itself may stimulate an effective immunity; in general, ES antigens have been more

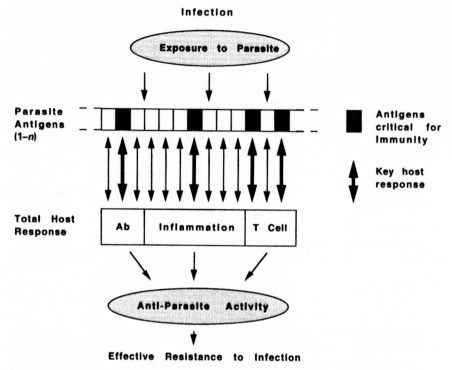

Fig. 3.1   Diagram illustrating that, although the host mounts a complex immune response to a complex array of parasite antigens, effective resistance depends upon key responses to a relatively small number of critical (functional) antigens.

effective. As a consequence, attention has been directed to antigens associated with, or released from, the living parasite as an essential requirement for successful immunization. Although such antigens represent a small subset of a parasite's total antigenic make-up, they are still complex mixtures. Nevertheless, it seems to be the case that certain of the components present are always immunodominant, and that these will contain the most important functional antigens (Fig. 3.1). Identification of these components and of the responses they elicit is therefore the central concern of immunoparasitology, and one in which rapid progress is being made.

Another area of difficulty with parasite antigens is the lack of information about their origin and biological function, or the ways in which they may be presented to the host (Fig. 3.2). Structural antigens, other than those which form components of surface membranes, will normally not be available to the host's immune system whilst the parasite is alive. This means that the host may recognize the antigens after the death of some of the parasites, but effector

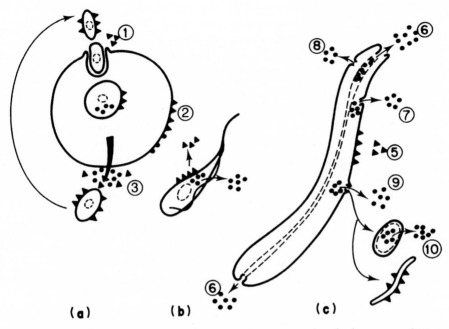

Fig. 3.2   Ways in which parasite antigens may be presented to the immune system of the host. ▼▼ represent surface (membrane) antigens; ●● represent internal antigens.
(a) Intracellular protozoa with extracellular invasive stages. (1) Antigens are present on the surface of invasive stages and may be released on entry. (2) Antigens from the intracellular stages appear on the membrane of the host cell and are released (3) when the cell ruptures to release the next generation of invasive stages.
(b) Extracellular protozoa. Antigens are presented on the surface. Both surface and internal antigens are released into the host tissues.
(c) Helminth parasites. (5) Antigens are present on the surface and are released into the host tissues. Internal antigens are released during feeding (6), excretion (7), moulting (8) and reproduction (9). Reproductive stages may continue to present antigens when retained in the body (10).

mechanisms may be unable to interact directly with them in living parasites. ES antigens may diffuse rapidly from the vicinity of the parasite. In consequence, recognition and effector interactions may occur at some distance from the parasite and thus be relatively ineffective. It should be clear from this that, in analysing immune responses that interact in a direct manner with living parasites and which can damage them, it is natural to look for antigens present at their surfaces, i.e. at the immediate host–parasite interface. As will be discussed later, however, not all immune responses that are effective against parasites act in this way. The generation of inflammatory changes plays an important role in protection against many parasites and in such cases ES antigens may initiate and focus the protective response.

With a number of intracellular protozoa, appearance of antigens at the host cell surface is an important requirement for the expression of immunity. Not only can these antigens be recognized by antibodies but, as discussed in Chapter 2, antigens expressed in this way are likely be bound to Class I self MHC molecules and can therefore become targets for cytotoxic T cells. Where MHC molecules are absent or poorly expressed, as on red blood cells, T cell-mediated cytotoxicity is correspondingly absent or much reduced. In certain cells (e.g. macrophages) parasite antigen may also be presented in the context of Class II MHC and be available for recognition by T helper cells. Although the intracellular location offers a considerable degree of protection, all intracellular protozoa must enter the extracellular environment at certain stages of their life cycle; when they do so they become vulnerable to direct immune attack. The antigens of protozoa that are permanently extracellular are always exposed to the immune system and parasite survival depends upon successful evasion of effector mechanisms. The ability of the African trypanosomes to change their surface antigens is perhaps the best-known example of such evasion strategies.

Helminths, as metazoan parasites, are larger and have a more complicated level of organization than protozoans. They therefore show a greater antigenic complexity. There are, however, sufficient examples in which particular antigens have been shown to have a major role in stimulating protective immunity to encourage the view that this complexity does not create an impossibly difficult situation for analysis. Platyhelminth worms, such as schistosomes, have bodies that are covered by a potentially vulnerable, plasma membrane-bounded surface. They release a wide variety of metabolic antigens, but the major antigens that are relevant to protective immunity are undoubtedly those located at the surface of the body. Nematodes, on the other hand, have a tough outer cuticle that certainly confers a degree of protection against immune attack. Despite earlier views to the contrary, it is now known that the cuticular surface is both antigenic, i.e. will bind antibodies, and is immunogenic, i.e. will stimulate immune responses. Antigen turnover and release into the surrounding environment has also been demonstrated. With certain species, and particularly with larval stages, immune attack directed against surface antigens can result in damage and death. The relevance of surface antigens to immunity is more apparent with tissue-dwelling worms, such as filarial nematodes, than it is with intestinal worms. In the latter, immunity seems more dependent upon antigens released during feeding, excretion, moulting and reproduction. In the case of feeding, inactivation of digestive enzymes by combination with antibody, though it occurs at some distance from the worm, may very effectively reduce viability and therefore lead to parasite elimination.

The location of antigens responsible for immunity to blood-feeding arthropods is more easily identified than in other parasites and the biological functions of the molecules concerned are more obvious. These antigens are almost entirely associated with feeding, particularly with secretion of saliva, and with the cells of the intestine. Interactions between immune effectors such as antibody and the intestine can prevent feeding and kill the parasite directly. Inflammatory responses to arthropod salivary antigens may also affect the parasite directly through release of mediators, but indirect consequences as non-specific as scratching can also provide effective protection.

3.4. Identification and measurement of immunity

Infection with parasites and exposure to their complex antigens generates a complex immune response. This can be measured using a variety of conventional immunological tests, such as those that determine levels or specificity of antibody and cellular activity. Responses measured in this way may, however, provide little or no correlation with protective immunity, indicating that many of the responses elicited by infection are irrelevant to the continued well-being of the parasite and confer no resistance on the host. They may represent responses to antigens that are not critical to survival, responses that do not interfere sufficiently with the functional integrity of antigens that are critical, or even diversionary responses that represent part of a parasite's evasion strategies. Protective immunity is best determined by parameters that measure parasite development, growth, reproduction and survival. Identification of the immune effector mechanisms responsible for this immunity can then come from three types of studies:

(a) correlation of particular responses with the *in vivo* expression of resistance, the approach most often used in analysis of anti-parasite immunity in humans;

(b) *in vitro* analysis of the ability of particular effectors to kill parasites;

(c) manipulations of host and parasite in experimental systems.

Correlative studies in humans have been most valuable when linked with detailed epidemiological data, or with *in vitro* analysis of potential effector mechanisms. Large-scale epidemiological studies of populations in regions endemic for particular parasites always identify groups of individuals whose

Table 3.1 *Mice as experimental hosts for immunoparasitological studies*

Strains available and their characteristics

Random-bred	Each individual is genetically different
Inbred	Individuals in a strain are genetically identical
Recombinant inbred	Series of strains derived from two inbred parental strains, with known chromosomal markers
Congenic	Inbred strains differing only at defined loci
MHC congenic	Inbred strains differing only at loci of the MHC
Mutant	Strains with known genetic mutations
Transgenic	Mice with defined genes inserted into the genome
Knock-out	Mice with defined genes deleted or inoperative

Reagents available

Antibodies to identify or deplete antibody isotypes
Antibodies to identify or deplete cells
Antibodies to identify or deplete cytokines
Antibodies to identify receptors and other cell components
Recombinant cytokines and growth factors
Cell lines for maintenance of parasites *in vitro*, production of hybridomas or secretion of cytokines
Gene sequences for immunoglobulins, cytokines, receptors

levels of infection differ markedly from the majority. Analyses of parasite-specific immune responses in these individuals can often provide clues to the responses causally associated with resistance or with susceptibility, and these can then be followed up using *in vitro* approaches.

Experimental studies in a wide variety of mammalian and avian hosts have provided the most detailed and most precise information we have about the operation of anti-parasite immune responses. Work with primates, for example, has been crucial to understanding immunity to malaria and schistosomes, and much of our knowledge of immunity to parasites of veterinary importance has come from work with cattle, sheep and chickens, but the majority of experimental work has involved the study of infections in laboratory rodents. The most commonly used rodents are mice, and much of the data referred to in subsequent chapters will relate to work with this host. Mice have the advantages of being genetically and immunologically well-defined, which allows very precise analysis of their immune responses to infection. It is possible to infect mice with a wide variety of parasites, including many of direct relevance to humans, and by studying these infections in random-bred, inbred, congenic, mutant and transgenic strains (Table 3.1) it is possible to pin-point resistance mechanisms and the genes that control them. Similarly, the enormous range of immunological reagents now available for mice can be used to identify and manipulate specific response components with great accuracy.

4

Intracellular protozoa
Survival within cells

4.1 Introduction

Several species of Protozoa live as intracellular parasites, occupying a variety
of cells within the body of the host. Included among these species are a
number that are of major medical and veterinary importance and responsible
for widespread disease (Table 4.1).

Life within cells poses certain problems for the protozoan, such as recognition and penetration of the correct cell, and transmission between hosts. At
the same time it confers many advantages, of which protection from potential
immune effectors is certainly one, and these advantages have, in evolution,
outweighed the difficulties inherent in this mode of life. Recognition and
penetration of suitable cells is dependent upon complex membrane interactions and will be discussed again later. Transmission between hosts is achieved
in many cases by arthropod vectors; in species that parasitize intestinal cells
(e.g. *Eimeria, Toxoplasma* in the final host) the normal mode of transmission is
via a resistant cyst. The manner in which the intracellular location influences
the immunological interactions between host and parasite is determined by
the nature of the cell occupied and will be discussed primarily in relation to
Plasmodium and *Leishmania.*

4.2 *Plasmodium* and malaria

Malaria ranks as one of the commonest and most important parasitic diseases.
It is estimated that some 1000 million people are at risk from infection and 200

Table 4.1 *Major intracellular protozoa of humans and domestic animals*

Parasite	Cell	Disease	Host
Babesia spp.	RBC	Piroplasmosis	Cattle
Plasmodium spp.	RBC, Hepatocytes	Malaria	Humans
Eimeria spp.	Intestine, Liver	Coccidiosis	Fowl, sheep, cattle
Toxoplasma gondii	Macrophages and many others	Toxoplasmosis	Humans
Leishmania spp.	Macrophages	Leishmaniasis (kala azar, oriental sore)	Humans
Trypanosoma cruzi	Macrophages, Muscle	Chagas' disease	Humans
Cryptosporidium	Enterocytes	Cryptosporidiosis	Humans

million are actually infected. In Africa, for example, infection leads to 1 million deaths annually, the majority of deaths occurring in children. The distribution of infection is determined by the vector arthropods, species of anopheline mosquitoes, and major foci occur in Africa, India, S.E. Asia, Central and S. America. Despite intensive programmes of vector control, which have been successful in some countries, there has been no significant global reduction in the extent of malarial infection. Indeed, in certain countries, after a period of control there have been serious resurgences, arising from breakdown of control measures and the emergence of insecticide and drug resistance. In India the annual number of cases in 1962 was 100 000, but this has risen again to 10 million in recent years.

Humans are the intermediate hosts of *Plasmodium*, the sexual stages of the life cycle taking place in the body of the mosquito (Fig. 4.1). Infection is initiated by the bite of an infected mosquito and the injection into the bloodstream of sporozoite stages contained in the insect's saliva. The sporozoite enters hepatocytes of the liver shortly after injection (possibly via Kupffer cells) undergoing growth and asexual reproduction (schizogony) to form a large pre-erythrocytic schizont. Rupture of the infected cell releases thousands of merozoites, which then penetrate red blood cells (RBC) to initiate the erythrocytic cycle. After a period of growth, during which the parasite passes through the ring and trophozoite stages, there is schizogony and production of the erythrocytic schizont, with division of cytoplasm and nucleus to form a relatively small number (32 or less) of merozoites. The infected cells then burst,

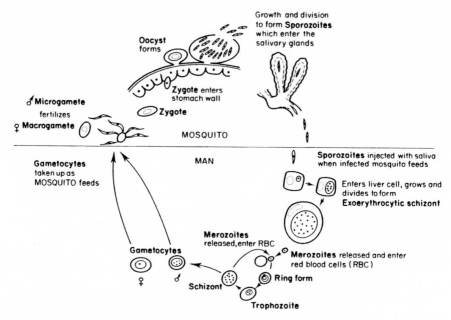

Fig. 4.1   Life cycle of *Plasmodium* in mosquito and man.

releasing merozoites, and repeated cycles of schizogony and RBC infection ensue, so that a high proportion of available RBC may become infected.

Growth of the parasite in the RBC is fuelled by intake and digestion of RBC cytoplasm. Haemoglobin is digested, to obtain amino acids, but the haem is stored in the form of an insoluble pigment. When the infected RBC bursts, pigment and other metabolic products are released into the circulation, inducing a number of changes in the host, of which fever is the most characteristic. In the human malarias there is synchrony of schizont formation and RBC rupture, and this is responsible for the regular and repeated bouts of fever that are almost diagnostic of the infection.

After several cycles of asexual division there is a switch to the sexual phase of the cycle. Merozoites invade RBC and produce male or female gametocytes rather than a further generation of schizonts. No further development can occur unless the cells containing gametocytes are taken up by a suitable mosquito. Production of male gametes and fertilization of the female gamete occurs in the stomach of the mosquito. The motile zygote then penetrates the stomach wall and forms an oocyst on the outer lining. Within the oocyst there is repeated division to produce large numbers of sporozoites and these eventually move anteriorly in the insect to enter the salivary gland.

In two of the species of human malarial parasites (*P. falciparum* and *P. malariae*) the duration of infection is determined by the duration of the asexual schizogonic phase in the RBC, and this may be very prolonged in the latter species. In *P. ovale* and *P. vivax* prolonged infections arise from relapses, i.e. from reinvasion of RBC by merozoites released from liver schizonts that have remained dormant.

The pathological consequences of malarial infection are primarily associated with the destruction of RBC, which can lead to anaemia and vascular collapse. In *falciparum* malaria, the most dangerous form, there is an additional primary cause of pathology. Cells containing schizonts are sequestered in capillaries of internal organs. If the brain is involved, the damage caused by blockage of capillaries can lead to the fatal condition known as cerebral malaria. Other pathological manifestations arise as secondary consequences of infection, for example the nephrosis that follows immune complex-mediated damage to the kidney.

4.3   Resistance and immunity to malaria

Plasmodium is an excellent example of a microparasite. Its small size enables it to exploit the specialized niche of the red blood cell. The ability to undergo repeated cycles of asexual reproduction means that, in theory, overwhelming infections can result from exposure to a single sporozoite. Transmission by an abundant and highly successful vector ensures that hosts are infected at an early age and may continue to be infected for much of their lifetime. These factors contribute to the high prevalence and widespread distribution of malaria, and explain why it is one of the world's most important infectious diseases. Nevertheless, human populations do co-exist with *Plasmodium* and individuals survive successfully even in highly endemic areas. There must, therefore, be at least the potential for significant resistance to infection. The mechanisms through which this resistance can be expressed and factors that reduce its effectiveness have been explored in great detail, through population studies, *in vitro* studies using human sera and cells, and a variety of experimental model systems in primate and rodent hosts (Table 4.2.). Five important concepts have emerged from this work:

- natural resistance
- species specificity
- stage specificity
- antigenic variation
- immune-suppression

Table 4.2 *Non-human species of* Plasmodium *used for experimental studies of malarial immunity in primate and rodent hosts*

Host	Parasite
Primate	*P. cynomolgi, P. inui, P. knowlesi*
Rodent	*P. berghei, P. chabaudi, P. vinckei, P. yoelii*

4.3.1   Natural resistance to malaria

Susceptibility and resistance to malaria in humans are influenced by several factors that have nothing to do with immunologically-mediated responses to infection and which may affect the parasite at distinct phases of its development. Entry of merozoites into the RBC depends upon a complex process initiated by interaction between receptors on the RBC membrane and receptor-binding proteins on the parasite surface. Certain genetic variants of RBC lack the appropriate receptor molecules and, as a result, are not penetrated by merozoites. This has been most clearly shown for merozoites of *P. vivax* and *P. knowlesi*, which are unable to enter cells lacking the Duffy blood group antigens. The absence of *vivax* malaria from areas of West Africa can be explained by the fact that the population contains a high proportion of Duffy negative individuals.

Natural resistance may also operate after merozoites have entered RBC. If the enviroment provided by the cell is unfavourable then the development of the parasite may be impaired and the host protected. The best-known example of such an effect is seen in individuals heterozygous for the gene responsible for sickle-cell haemoglobin (HbS), in which there is a substitution of valine for glutamic acid in the β chain of the molecule. These individuals have substantial protection against *falciparum* malaria. In The Gambia, for example, carriers of the HbS gene were more than 90% protected against severe clinical disease. Although *P. falciparum* can develop perfectly well in HbS-containing cells, so there is no protection against infection as such, when infected cells are exposed to low oxygen tensions during sequestration in capillaries, there is leakage of potassium from the cell and the parasites are killed, reducing the pathologic complications associated with this phase. In the homozygous state the sickle cell condition is invariably fatal, and there is no doubt that the persistence and high frequency of the HbS gene in malarial endemic areas has resulted from the selection pressure the parasite has exerted on survival of the pre-reproductive population. A number of other haemoglobin and RBC variants (e.g. the α and β thalassaemias and G6PD deficiency) have also been maintained in populations by this selection pressure.

4.3.2   Acquired immunity to malaria

Despite the fact that infection exposes the immune system to a very considerable antigenic challenge, there is in humans at best only an incomplete immunity to the parasite, and this wanes rapidly in the absence of reinfection. That some degree of immunity does develop can be deduced from the fact that in endemic areas adults survive with intermittent parasitaemia (presence of parasites in the blood) and little serious pathology, whereas children are heavily infected and suffer extensively from clinical disease. Babies under the age of three months are relatively resistant to infection and this is attributable both to transfer of maternal antibody and to the continuing presence of foetal haemoglobin. More direct evidence for immunity has been gained from *in vitro* experiments, e.g. by studying the effects of serum taken from patients upon the growth, division and survival of malaria parasite in cultures of infected RBC. Why then is immunity not more effective? Fig. 4.2 summarizes some of the explanations that have been proposed. These include:

- Malaria parasites show considerable *antigenic diversity* and *variation* – between species, between strains, between stages and during the course of infection – immunity to one species, strain, stage or variant gives little protection against others.
- Parasites avoid immunity by '*hiding*' inside cells or by being *sequestered* away from organs such as the liver and spleen that are rich in phagocytic cells.
- Infection stimulates T-independent immunity and there is *little T cell memory*.
- Parasites *misdirect* or *suppress* the immune response.

4.3.3   Species specificity of immunity

There is good clinical evidence, and some supporting experimental data, that in humans immunity against one species of malaria does not confer protection against the others; this has also been confirmed using human malarial infections in primates. Primate malarias themselves generate species-specific immunity and there is only limited cross-immunity between the four rodent

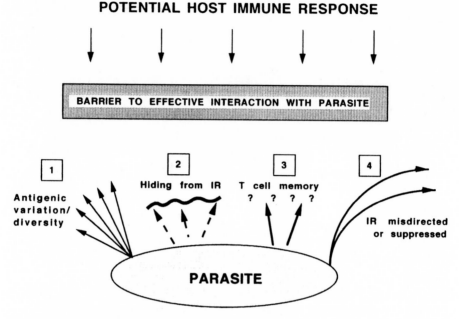

Fig. 4.2 Possible explanations of the relative ineffectiveness of immunity to malaria infections in humans.

malarias most commonly used in experimental studies. (It is interesting, however, that there does seem to be cross-protection between human and rodent malarias, e.g. *P. falciparum* and *P. berghei*.) Two important conclusions stem from these observations. Firstly, that there are species-specific, protection-inducing antigens and, secondly, that protective immunity involves recognition of these specific antigens by immunologically specific effector mechanisms. Although these conclusions may seem obvious, they lead to the inference that cross-reacting antigens, and effectors elicited by these antigens, play little or no part in protective immunity, which allows them to be eliminated as possible candidates for vaccine-induced protection.

4.3.4   Stage specificity of immunity

Acquired immunity operates specifically against particular stages of the life cycle. For example, immunity generated experimentally by exposure only to the erythrocytic stages has no effect upon the initial pre-erythrocytic stages of infection and *vice versa*. This again indicates that there are restricted antigens, in this case stage-specific antigens. It is also clear that there must be significant differences between the immune mechanisms that could be effective against the various extracellular stages of the cycle, the sporozoites and merozoites, and those effective against the intracellular stages in the hepatocyte and the red cell. Each will be considered separately.

4.3.4.1   Pre-erythrocytic stages

(a) Sporozoites. In naturally acquired infections sporozoites are present in the blood for only about 45 minutes and appear to generate little protective immunity, even though anti-sporozoite antibodies are present in the sera of older children and adults from endemic areas. Nevertheless, it is well established that the sporozoite possesses antigens capable of eliciting protective immunity and that these are surface molecules. For example, exposure to radiation-attenuated sporozoites has been used to generate an effective resistance against homologous challenge both in humans and in mice. Immunization results in the formation of antibodies that react primarily with antigens at the sporozoite surface and incubation in immune serum *in vitro* results in the formation of a circumsporozoite precipitate (Fig. 4.3). Incubated sporozoites lose their infectivity probably because their ability to recognize and invade liver cells has been lost, although, clearly, antibody coating and immobilization would also facilitate uptake by phagocytic cells. *In vitro* studies have confirmed that antibody blocks the ability of sporozoites to invade hepatocytes, and it has been suggested that cross-linking of surface molecules may reduce sporozoite motility.

Much of the sporozoite surface is covered by a single protein – the circumsporozoite protein (CSP) – the structure, immunology and molecular biology of which is known in great detail for several species. The genes concerned have been cloned and the amino acid sequence of the molecules elucidated. CSP peptides can now be produced synthetically or by recombinant techniques. In *P. falciparum* the CSP has a MW of about 60 kDa and is characterized by some 45 repeats of the amino acid sequence asparagine-alanine-asparagine-proline [NANP] (Fig. 4.4). These NANP repeats form the dominant epitope recognized by anti-CSP antibodies. Antibody production is T cell dependent, epitopes recognized by T cells being present in the repeat region and in the flanking sequences. The T cell epitopes show polymorphism within species. CSP contains a conserved sequence that may facilitate sporozoite invasion, by binding to proteoglycans on the hepatocyte surface.

Early studies using *P. berghei* in the mouse model showed that prior injection of anti-CSP monoclonal antibodies gave a very high degree of protection against homologous challenge, even when as little as 10 μg of antibody was

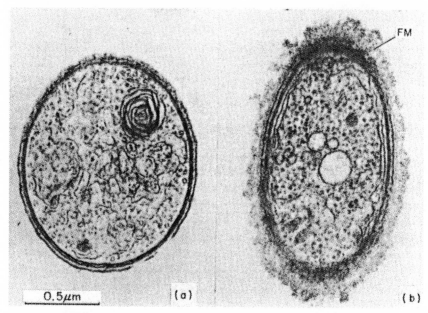

Fig. 4.3   Surface antigens of malarial sporozoites. Electron micrographs through (a) sporozoite of *Plasmodium berghei* incubated in normal mouse serum; (b) sporozoite of *P. berghei* incubated in immune mouse serum. The surface is covered by a thick coat of fibrillar material (FM). (Photographs from Nussenzweig *et al.*, 1978, in *Rodent Malaria*, eds. Killick-Kendrick & Peters, p. 248, Academic Press, New York, by permission of the authors and publishers.)

used, leading to the conclusion that antisporozoite immunity was primarily antibody-mediated. However, mice treated from birth with anti-μ chain anti-sera, and incapable of antibody formation, could still respond to vaccination with irradiated sporozoites and resist a challenge with fully infective sporo-zoites, leading to the opposite conclusion, that immunity had a cellular basis. This conclusion was strengthened by the fact that T cell depleted mice could not be protected.

(b) Intra-cellular stages. There is now good evidence that protective immunity against the pre-erythrocytic stage of malaria can be mediated *in vitro* through cytotoxic mechanisms targeted against parasite antigens expressed on the surface of infected hepatocytes. These include residual sporozoite antigens (CSP and the second sporozoite surface protein – SSP2) as well as new anti-gens such as the liver stage antigens LSA-1 and 2. CD8[+] cytotoxic T cells spe-cific for such antigens have been found in individuals naturally infected with

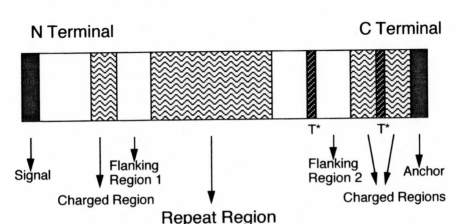

N Terminal   C Terminal

Signal   Flanking   Flanking   Anchor
         Region 1   Region 2
Charged Region   Charged Regions
         Repeat Region

T*   T*

P. falciparum: NANP (major x 40) / NVDP (minor)
P. vivax: ACQPAGRA / DCQPAGRA

Fig. 4.4   The circumsporozoite (CS) protein of *Plasmodium falciparum*, showing the major structural and functional regions. The amino acid sequences of the central repeats in *P. falciparum* and *P. vivax* are also given. T*, T cell epitope. (Modified from Mitchell, 1989, *Parasitology* **98**, S29.)

P. falciparum, and both CD8$^+$ and CD4$^+$ cytotoxic T cells have been reported in immunized humans and mice, but the *in vivo* role of these cells remains unclear. It has also been proposed that a number of cytokines (IL-1, IL-6 IFN-γ, TNF) may be important in bringing about intracellular parasite destruction, either directly or via production of short-range mediators such as nitric oxide (Fig. 4.5).

Sporozoite antigens have been considered as one of the major candidate antigens for an anti-malarial vaccine (see page 182).

4.3.4.2   Erythrocytic immunity

(a) Merozoites These are the second extracellular phase of the cycle, and must survive in the face of developing immunity if the infection is to continue. The degree of immunity against the erythrocytic cycle varies considerably between species. In humans, for example, immunity develops only slowly, whereas in mice infected with *P. yoelii* a sterile immunity, i.e. one where no parasites remain, may develop quite rapidly. Irradiated merozoites have been used successfully to vaccinate primates against *P. knowlesi* and merozoite antigens used to vaccinate *Aotus* monkeys against *P. falciparum*.

The merozoite is a vulnerable target for immune effector mechanisms, particularly antibodies. Immunity can be transferred passively, and such transfer

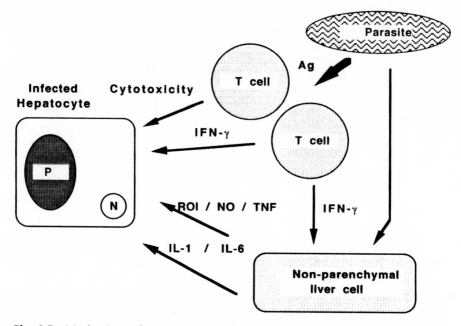

Fig. 4.5 Mechanisms of immune attack directed against the exoerythrocytic stages of malaria parasites in hepatocytes. Immunity involves T cells (both CD4⁺ and CD8⁺) and non-parenchymal cells and is effected through a complex network of cytokines (IFN-γ, IL-1, IL-6, TNF) as well as mediators such as nitric oxide (NO) and reactive oxygen intermediates (ROI). Ag, parasite antigen; N, hepatocyte nucleus; P, parasite.

has been achieved in humans as well as in several rodent models; conversely infections are more severe in animals rendered incapable of antibody production. The action of antibody can be studied *in vitro*, using continuous culture of infected RBC, in which the regular cycles of growth, schizogony, merozoite release and cell invasion occur quite normally. Although the *in vivo* relevance of such studies is debatable, those made with *P. falciparum* and *P. knowlesi* have shown that antibody can block merozoite entry into RBC (Fig. 4.6), either by causing parasite agglutination or, possibly, as a consequence of antibody binding to surface molecules that are needed for interaction with RBC receptors. Blocking can be achieved with the F(ab)₂ fragment of IgG₁ antibody (i.e. with the divalent antigen-binding fragment) and is complement independent. Although combination of antibody with free merozoites should lead to complement-mediated lysis there is little evidence for this.

Merozoites express a number of antigens at their surface and release antigens from organelles such as the rhoptries and micronemes during invasion of the RBC (Fig. 4.7). Several have now been isolated and sequenced and given a

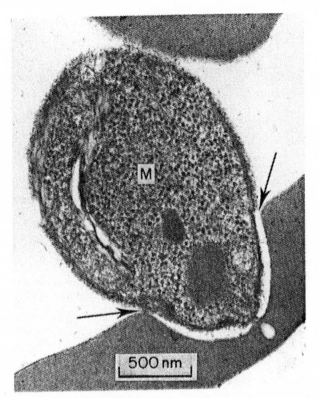

Fig. 4.6 Entry of merozoite of *Plasmodium knowlesi* into host red blood cell, showing invagination of cell surface and attachments between rim of invagination and merozoite surface (arrowed). M, merozoite. (Photograph from Bannister *et al.*, 1975, *Parasitology*, **71**, 483, by permission of the authors and publishers.)

confusing variety of designations. An important group of surface antigens is derived from a large MW protein molecule – MSP1 (merozoite surface protein 1) – itself derived from a larger MW precursor. In *P. falciparum* MSP1 is also referred to as Pf200 or gp195. MSP1 may play a role in interaction with the red cell membrane. Monoclonal antibodies directed against epitopes on MSP1 inhibit RBC invasion, and purified MSP1 from a number of species has been used to elicit protective immunity. MSPl is processed into a series of smaller molecules that appear at the merozoite surface at about the time of shizont rupture. In *P. falciparum*, for example, MSP1 is cleaved into fragments ranging between about 19 and 83 kDa. The structure of MSP1 from *P. falciparum* is known in detail and it is clear that there are significant polymorphisms in the molecule. Merozoites also express a second major surface protein (MSP2 – Pf 48–53), the two MSPs being encoded on different chromosomes.

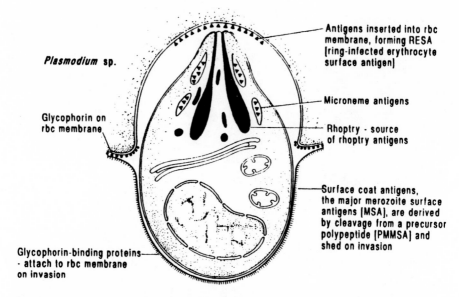

Fig. 4.7 Merozoite invasion of the red blood cell, showing the antigens present at this stage of the life cycle. (Adapted with permission from Wyler, 1990, *Modern Parasite Biology. Cellular, Immunological and Molecular Aspects*, W.H. Freeman, New York.)

Rhoptry antigens (RAP – rhoptry associated protein) are thought to play an important role in RBC invasion, and antibodies to these are known to provide protection against challenge infections. Other important merozoite-related antigens are also associated with RBC invasion, e.g. the glycophorin-binding proteins and the erythrocyte-binding antigens.

4.3.4.3 Schizont-infected RBC It has been assumed that the schizogonic stages are protected by virtue of their intracellular position (Fig. 4.8) and there is some experimental evidence to support this assumption (although it may have to be modified now there is good evidence for the existence of a duct connecting the parasitophorous vacuole to the exterior of the cell). In addition, the RBC membrane expresses very few if any MHC molecules and, as these molecules are essential for expression of the cytolytic function of effector T cells, there can be no lymphocyte-mediated cytotoxicity against infected RBC. However, the host is capable of agglutinating parasite-infected RBC and also of killing parasites within them by a variety of means. In certain primate and rodent models peak parasitaemia is associated with the appearance of 'crisis-forms', the parasite becoming compact and densely-staining and

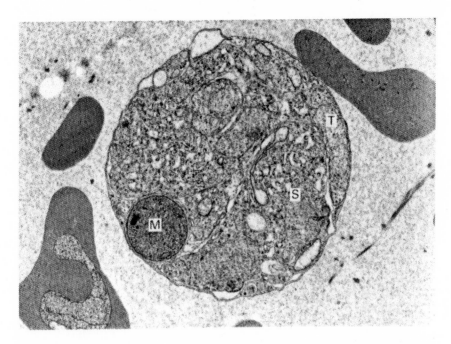

Fig. 4.8   Red blood cell multiply infected with stages of *Plasmodium yoelii*. M, recently invaded merozoite; S, schizont; T, trophozoite. (Photograph by courtesy of Professor R.E. Sinden.)

showing signs of degeneration. It has been suggested that crisis forms result from the release from macrophages, neutrophils and possibly natural killer cells, of short-range cytotoxic factors that can cross the RBC membrane and interact directly with the parasite. Candidate factors include reactive oxygen metabolites, nitric oxide and cytokines such as TNF. The nature of these factors, which are rapidly inactivated, suggests that they are likely to be effective only when parasitized RBC (or free merozoites) are in close contact with host cells, for example when passing through the spleen or liver. It is possible that effector molecules may also be able to affect the intracellular stages after passing from the exterior through the connecting duct to the parasitophorous vacuole.

Antigens from the intracellular stages of the parasite are present on infected RBC. Some of these derive from the initial invasion, others are transported from the parasitophorous vacuole (Fig. 4.9). In *P. falciparum*, for example, merozoite invasion results in the appearance of a 155 kDa molecule (RESA – ring-infected erythrocyte surface antigen or Pf155) derived from the micronemes and associated with the RBC cytoskeleton.

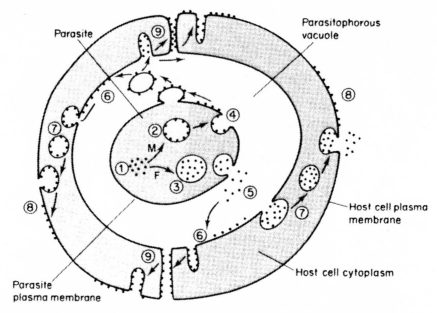

Fig. 4.9   Models of the transport of protein molecules to the surface of malaria-infected red blood cells. Two pathways are shown, one (M) involving membrane-bound molecules, the other (F) involving free molecules. Protein is synthesized in the parasite cytoplasm (1) and translocated into the endopasmic reticulum (2) or taken up into into vesicles (3). After fusion with the parasite plasma membrane (4) membrane-bound proteins are carried in blebs, which then fuse with the membrane of the parasitophorous vacuole (6). Proteins are transported in vesicles across the red cell cytoplasm (7) to be expressed at the surface of the host cell plasma membrane (8). Free molecules (5) may be endocytosed (7) and carried to the host cell surface in vesicles to be retained on the plasma membrane (8) or be released; they may escape through temporary ducts (9) or be released from the parasitophorous vacuole when the cell ruptures. (Based upon Howard, 1988, in *The Biology of Parasitism*, eds. Englund & Sher, p. 111, Alan R. Liss, New York.)

Recombinant peptides from this molecule have been used to raise protective immune responses in *Aotus* monkeys. At a later stage, RBC infected with certain species may express parasite-derived proteins that are associated with changes in RBC morphology and the appearance of knob-like extensions of the cell membrane. A number of such proteins has been identified, including the histidine-rich knob-associated proteins (HRP1/2), the mature-infected erythrocyte surface antigen (MESA) and the erythrocyte-membrane proteins (EMP1/2). Expression of these proteins (EMP being the most important) causes the cell to adhere to a variety of receptors on the endothelial cells of capillaries – the process of sequestration. These receptors, which are themselves upregulated during infection, include CD36; intercellular adhesion molecule 1 (ICAM-1); E-selectin; vascular cell adhesion molecule 1 (VCAM-1) and thrombospondin. Infected cells also express antigens known as rosettins, which cause uninfected RBC to adhere in rosettes.

Sequestration of trophozoite and schizont-infected cells of *P. falciparum* in the vessels of internal organs such as the brain probably protects the parasites during this critical developmental stage, by preventing their passage through phagocytic organs such as the spleen, but is a major factor in the development of severe host pathology. The presence of antibodies to these proteins helps to prevent sequestration, although EMP can be antigenically variable. It is probable that other parasite-derived molecules are inserted into the host cell membrane, but these have yet to be fully characterized. One that has is a molecule thought to act as a transferrin receptor, and to be involved in transporting iron to the parasite from the host's plasma.

In addition to these surface antigens, there is good evidence that malarial parasites secrete antigens that are released when infected cells rupture (Fig. 4.10). These include the glycophorin-binding proteins (GPB), the erythrocyte-binding antigens (EBA), heat shock proteins (HSP), and the S-antigens, an antigenically diverse group of proteins that can be detected in the sera. S antigens may elicit protective immunity, but their unusual structure, with extensive and cross-reactive tandem repeat sequences, may also contribute to the generation of low-affinity, and therefore poorly protective antibody responses (see below). The high MW *falciparum* interspersed repeat antigen (FIRA), which is released at RBC lysis, may also act in this way. Rupture also releases toxic components that trigger episodes of fever (see section 4.6.2).

4.3.4.4 Gametocytes
Antibodies against gametocyte antigens are known to occur, for example, in humans infected with *P. falciparum* and *P. vivax*, but they have no effect on the sexual stages while they are in the human. However, antibodies in the blood taken up by mosquitoes as they feed can influence the subsequent sexual phase of the cycle in the insect. These antigens, and others present on the zygote, have been considered as candidates for use in transmission-blocking vaccines (see page 184).

Fig. 4.11 summarizes the relationship of some of the major parasite antigens to the life cycle stages concerned.

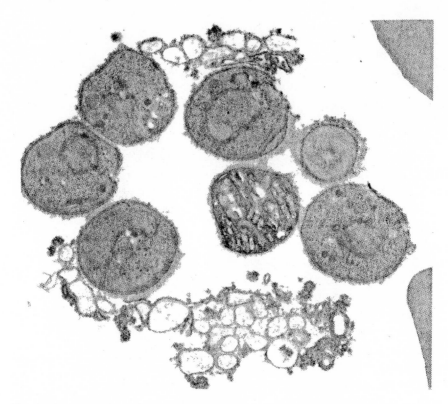

Fig. 4.10   Merozoites of *Plasmodium falciparum* from ruptured red blood cell, showing material released during this process. (Photograph by courtesy of Professor Dr H. Melhorn.)

4.4   Components of protective responses

The relative importance of individual components of the immune response in protection against malarial parasites is very difficult to assess. It is clear that immunity is T cell-dependent, and that antibodies can play a significant role, e.g. in blocking penetration of cells, in interfering with parasite nutrition, in agglutination/opsonization of invasive stages and in opsonization of infected RBC. The effectiveness of antibodies was demonstrated in the now classical experiment of Cohen and colleagues in 1961, who showed that transfer of immunoglobulins from adults immune to malaria into infected children was followed by a rapid fall in parasitaemia. Passive transfer of immunity in humans has also been described more recently (from African into Thai patients), IgG1 and IgG3 being identified as protective isotypes. Successful

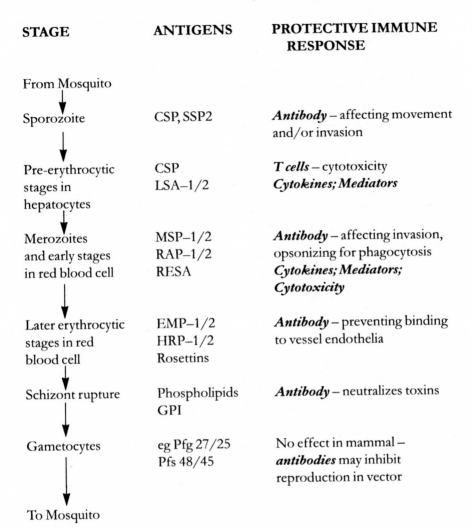

STAGE	ANTIGENS	PROTECTIVE IMMUNE RESPONSE
From Mosquito ↓		
Sporozoite ↓	CSP, SSP2	*Antibody* – affecting movement and/or invasion
Pre-erythrocytic stages in hepatocytes ↓	CSP LSA–1/2	*T cells* – cytotoxicity *Cytokines; Mediators*
Merozoites and early stages in red blood cell ↓	MSP–1/2 RAP–1/2 RESA	*Antibody* – affecting invasion, opsonizing for phagocytosis *Cytokines; Mediators; Cytotoxicity*
Later erythrocytic stages in red blood cell ↓	EMP–1/2 HRP–1/2 Rosettins	*Antibody* – preventing binding to vessel endothelia
Schizont rupture ↓	Phospholipids GPI	*Antibody* – neutralizes toxins
Gametocytes ↓	eg Pfg 27/25 Pfs 48/45	No effect in mammal – *antibodies* may inhibit reproduction in vector
To Mosquito		

Fig. 4.11 Summary of antigens released by stages in the malarial life cycle and protective immune responses associated with them (based on *Plasmodium falciparum*) .
CSP, circumsporozoite protein; EMP-1/2, erythrocyte membrane proteins 1/2; GPI, glycosylphosphatidylinositol; HRP-1/2, histidine-rich proteins 1/2; LSA-1/2, liver stage antigens 1/2; MSP-1/2, merozoite surface proteins 1/2; Pfs, *P. falciparum* sexual stage antigens; RAP-1/2, rhoptry-associated proteins 1/2; RESA, ring-infected erythrocyte surface antigen; SSP2, second sporozoite surface protein.

passive transfer of immunity, particularly against the erythrocytic stages of infection, has also been achieved in many experimental model systems, IgG isotype antibodies again being most effective. Transfer can result in a delayed appearance of patent parasitaemia, a reduced peak parasitaemia, an earlier resolution of infection, or all three (Fig. 4.12a). There is evidence that antibodies are maximally effective when other host responses are also operative, for example when phagocytic activity is increased. Conversely, antibody is less effective in splenectomized animals. It is assumed that phagocytic cells in the liver and spleen play an important role in clearing parasitized RBC from the circulation, but there is some controversy over how this is achieved. Opsonization of cells would appear to be essential, but increased uptake of infected RBC may also reflect parasite-induced altered shape and deformability of the cell and the large increase in spleen size (splenomegaly) that accompanies infection.

Experimental studies have shown that immunity can be transferred adoptively with lymphocytes and in some cases more effectively than by passive transfer of serum. Despite variations between systems, transfer of B cells, or mixed populations of B and T cells, has given the best results. These data reinforce the importance of antibodies in protective immunity. Nevertheless there is now substantial evidence that T cells also have a protective role that is distinct from that concerned with B cell help. Thus, B cell-deficient mice can express immunity to infection, the precise nature of the protection (e.g. against primary or against challenge infection) depending upon the malarial species. For example, initial erythrocytic infections with *P. yoelii* are primarily controlled by antibody and B cell-deficient mice will succumb unless treated with an antimalarial drug. However, if protected in this way, such mice can resist a challenge infection, even though they have no detectable antibodies. In contrast, initial infections with *P. chabaudi adami* are controlled by T cells, and B cell-deficient mice clear infections without chemotherapy.

T cells can transfer immunity very effectively, as shown, for example, by adoptive transfer of T cell clones into nude mice. CD4+ T cells are concerned primarily with immunity against the erythrocytic stages of infection (Fig. 4.12b), CD8+ T cells with immunity against the pre-erythrocytic stages. Deletion of these cell populations, by treating mice with appropriate anti-CD4 or anti-CD8 monoclonal antibodies, removes the ability to control blood- or sporozoite-stage infections respectively. However, these subset-specific activities are not absolute, both CD4+ and CD8+ clones specific for epitopes of the CSP can mediate activity against the pre-erythrocytic stages, and deletion of CD8 cells may interfere with clearance of the erythrocytic stages.

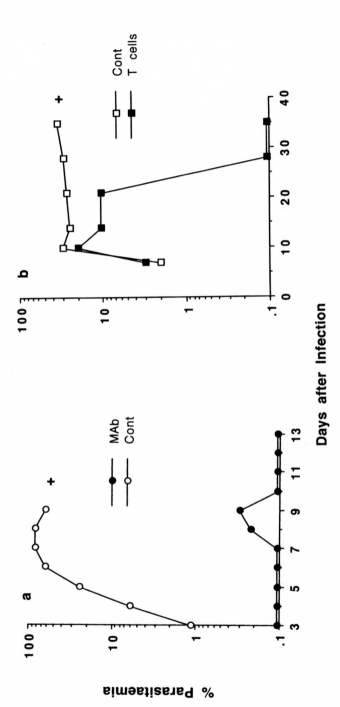

Fig. 4.12 Transfer of immunity against malaria infections in mice.
(a) *Plasmodium yoelii* in BALB/c mice. Experimental mice were given three injections of an anti-merozoite monoclonal antibody (MAb) before infection. Controls were given ascites fluid from the myeloma used. (Redrawn from Majarian *et al.*, 1984, *Journal of Immunology* **132**, 3131.)
(b) *Plasmodium chabaudi adami* in BALB/c *nu/nu* (T cell deficient) mice. Experimental mice were given cloned immune T cells before infection. Controls received no cells. (Redrawn from Brake *et al.*, 1988, *Journal of Immunology* **140**, 1989.)
Cont, controls; +, all mice died after this time point.

Table 4.3 *Characteristics of CD4 T cell lines that can transfer immunity against* Plasmodium chabaudi chabaudi *in mice*

	Derivation		Transfer		Help for	CK production			Subset
Line	Infection	Days p.i.	T+	T–	Ab	IFNγ	IL-2	IL-4	involved
775	Primary	16	+	+	–	+	–	–	Th1
779	Primary	20	+	+	–	+	+	–	Th1
737	Secondary	7	+	–	+	–	–	+	Th2
723	Tertiary	7	+	–	+	–	–	+	Th2

Notes:
T cell lines were established *in vitro* from spleen cells of mice given one, two or three infections. Cells were transferred into normal (T+) or T-deprived (T–) mice before infection. In the latter, Th 2 cells transferred immunity only if B cells were also given.
Source: (Data from Taylor-Robinson & Phillips, 1993, *Parasite Immunology*, **15**, 301.)

CD4$^+$ cells appear to play multiple roles in immunity to malaria. They provide help for production of antibody and for generation of cytotoxic cells, and release cytokines involved in activation of macrophages and related cells. Experiments with *P. chabaudi chabaudi* have shown that the balance of CD4$^+$ subset activity may change as infection progresses (Table 4.3) although it is not clear how typical this is of malaria infections in general. With this species, recovery from the first peak of parasitaemia was associated with Th1 and IFN-γ activity, presumably mediated via activated macrophages, whereas Th2 cells predominated during the later chronic phase and mediated eventual parasite elimination, probably via anti-parasite antibody. CD8$^+$ cells have a cytotoxic function against infected hepatocytes, and this involves the activity of a complex of cytokines (IFN-γ, IL-1, TNF-α and IL-6) that promote killing mechanisms, in which production of nitric oxide plays a major role.

The question of T cell memory in malaria infections has received a great deal of attention, given the importance of T cells in immunity and the evidence that protection is short-lived. One complicating factor in attempting to determine the extent of memory in humans is the fact that T cells from non-infected individuals can respond to parasite antigens, presumably as a result of prior exposure to cross-reacting environmental antigens. This difficulty has been overcome by testing T cell responses against a series of synthetic peptides from the CSP, which allows identification of epitopes only recognized by individuals who have been exposed to infection. The data obtained show that memory to *P. vivax* appears to persist for many years, probably at least 50, whereas that to *P. falciparum* is short-lived, waning after two years. This difference may be related to the fact that *P. vivax* forms persistent liver stages (hypnozoites) from which blood stage relapses originate, and which presumably provide long-term antigenic stimulation.

Most studies on T cell immunity in malaria have focused on the cells bearing the conventional alpha–beta T cell receptor. Recent data suggest that malaria infections are associated with an increase in number of gamma–delta receptor-bearing cells, and the possibility therefore exists that these cells, which are not MHC restricted, may also have a protective role.

4.5 Malarial antigens

4.5.1 Antigen cross-reactivity

One of the striking characteristics of *Plasmodium* antigens is their extensive cross-reactivity. This occurs at all levels:

(*a*) between different epitopes within a single antigen;
(*b*) between epitopes of different antigens of a given stage;
(*c*) between epitopes of the same antigen in different stages;
(*d*) between epitopes of different antigens in different stages.

This cross-reactivity is to a large extent associated with the unusually high frequency of repeat sequences in malaria antigens and exerts powerful influences on the type and effectiveness of immune responses. In particular it has been proposed that it contributes to an overstimulation of B cells, preventing antibody affinity maturation and so diverting the immune response into a 'smokescreen' of low affinity antibodies that have only limited protective value.

4.5.2 Antigenic variation

Malarial infections, particularly those established in natural hosts, are characterized by chronicity, with patent parasitaemia recurring at intervals after the initial acute parasitaemia has been controlled. In some species, notably *P. vivax*, recurrent parasitaemias are true relapses, the new erythrocytic cycles being initiated by merozoites released from latent exo-erythrocytic schizonts in the liver. In other species recurrent parasitaemias are recrudescences, derived from the multiplication of pre-existing erythrocytic stages. In laboratory models there is evidence that recrudescence follows a decline in effective immunity, but a more significant factor is the existence of antigenic variation, each recurrent parasitaemia representing a population that is antigenically distinct from the preceeding population and thus less well controlled by the host. Such phenotypic variation was first described for *P. berghei* in mice and is known to occur also in *P. cynomolgi, P. chabaudi chabaudi, P. fragile* and *P. knowlesi*. In the latter three species variation is seen at the level of parasite-derived variant molecules expressed at the RBC surface.

In *P. falciparum*, the emphasis in terms of antigenic variation is both on phenotypic change, as seen *in vitro*, and on genotypic diversity. The detailed molecular understanding we now have about this species shows that there is diversity at all levels, from the chromosomal to the individual gene. Diversity in the structure and immunological properties of major antigens is also well documented. This extensive polymorphism explains many of the characteristic features of infection in endemic areas. Although age at initial infection may be a contributory factor, the slow rate of acquisition of resistance under conditions of natural infection is now seen primarily as a consequence of exposure to antigenically diverse populations of the parasite, against each of which strain-specific responses have to be made. Polymorphism also creates some problems for vaccination strategies. In *P. falciparum*, for example, the immunodominant B cell epitopes created by the central repetitive sequences of the CSP are relatively well conserved, but there is considerable polymorphism in the flanking T cell epitopes required for initiation of responses.

4.6   Immunomodulation and immunopathology

4.6.1   Immunomodulation

It has been known for many years that malarial infections cause depressed responsiveness to unrelated antigens (Fig. 4.13). In humans this was shown originally by measuring the antibody responses of infected and uninfected children to vaccination with tetanus toxoid. Later studies extended these observations to a number of other vaccines, such as typhoid and polio, but demonstrated that cell-mediated responses to BCG, *Candida* and streptococcal antigens were normal. Paradoxically, malarial infection is also associated with increased synthesis of immunoglobulin, particularly IgM and IgG.

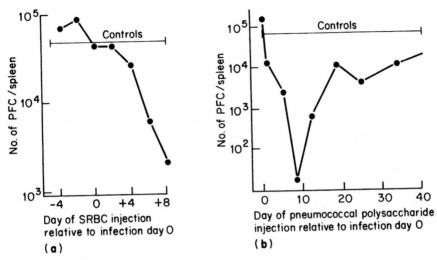

Fig. 4.13 Depression of immune responses in malaria infections.
(a). Response to sheep red blood cells (SRBC) in mice infected with *Plasmodium berghei*. (Redrawn from Liew *et al.*, 1979, *Immunology* **37**, 35.)
(b). Response to type III pneumococcal polysaccharide in mice infected with *Plasmodium yoelii*. (Redrawn from McBride *et al.*, 1977, *Immunology* **32**, 635.)
SRBC is a T cell dependent antigen, pneumococcal polysaccharide is T cell independent. Responses were measured by counting the numbers of direct plaque-forming cells in the spleen after 5 (a) or 4 (b) days.

Clinically immune West Africans produce about seven times as much IgG per day as do uninfected Europeans, and the level of synthesis falls markedly after anti-malarial drug treatment. It is thought that these increased levels of Ig result from polyclonal B cell activation, and that much of the Ig is not specific for malarial antigens. Direct evidence has been obtained that a mitogenic material is released from parasites during *in vitro* culture. If such a mitogen is also released *in vivo* it would contribute to this polyclonal activation.

Immunosuppression may also reflect altered immunoregulatory processes. Interference with antigen presentation by macrophages, and with normal patterns of cytokine production, have been demonstrated. Infection is associated with profound changes in the structure of lymphoid organs and in populations of circulating lymphocytes, and these too may disrupt normal immune function.

4.6.2   Immunopathology

Malaria infections are associated with a wide variety of pathological symptoms, of which recurrent fever is the most characteristic. It is likely that fever is caused, in part, by release of cytokines (e.g. IL-1 and TNF) from phagocytic cells as they take up material released when parasitized RBC rupture. Parasite-derived phospholipids, especially the glycosylphosphatidylinositol (GPI) anchor component of membrane proteins, also trigger fever. These molecules have been shown to act on macrophages in much the same way as bacterial endotoxins, i.e. they induce activation and cytokine release. However, many other aspects of malarial disease also involve components of the immune and inflammatory systems. Attention has focused recently on the role of TNF in the immunopathology associated with *falciparum* malaria. TNF is pleiotropic in its effects on the body and many correlates of malarial disease can be mimicked by injection of TNF, e.g. the neurological symptoms and the severe anaemia. TNF production is further potentiated in the presence of cytokines such as IFN-γ, and its activity is likewise mediated through complex pathways in which other cytokines and effector substances such as nitric oxide are involved. The glomerulonephritis found in both acute and chronic malaria, typically in *P. malariae* infection, is a further form of immunopathology. The cause of the kidney lesions is the deposition of immune complexes in the glomeruli.

It is important to remember that some of the pathological symptoms associated with infection are the consequences of responses that can have beneficial effects. The raised body temperatures experienced during malarial fevers, for example, also help to destroy parasites. TNF can be protective, by helping to kill parasites, as well as cause pathology – it is the balance of cytokine activity that is important.

4.7   Genetic control of immunity

The pathological consequences of malaria infection have exerted strong selection pressures over long periods of time. As a result the frequencies of certain genes, notably those associated with the haemoglobin and other red cell-related variants that influence natural resistance to infection, are significantly increased in endemic areas. Recent studies have also shown significant correlations between MHC genes and resistance to malarial disease. These correlations involve both Class I and Class II genes and have been detected by comparative HLA-typing studies on individuals with severe or mild *falciparum* malaria and those with other infections. The Class I antigen B53 was linked with approximately 40% protection against cerebral malaria and malarial anaemia and the Class II antigen DRB1 1302-DQB1 with protection against anaemia only. The frequency of these alleles in endemic areas such as The Gambia is very much higher than in non-exposed populations, indeed HLA B53 is common in sub-Saharan Africa but relatively rare elsewhere. The Class I association may reflect cytotoxic T cell-mediated protective immunity against the pre-erythrocytic liver stages of infection. Direct analysis of the parasite peptides bound by the B53 molecule have shown these to be derived from the liver stage antigen (LSA-1) expressed on the surface of infected hepatocytes.

4.8   Leishmania and leishmaniasis

All species of *Leishmania* are parasitic in cells of the mononuclear phagocytic series, primarily macrophages, cells designed to phagocytose and destroy invading organisms and other foreign bodies. This apparently paradoxical choice of host cell is shared with many bacterial pathogens and with a number of other protozoa, such as *Trypanosoma cruzi* and *Toxoplasma gondii*. All are capable of prolonged intracellular survival and reproduction, and exhibit a variety of survival strategies.

The species of *Leishmania* that can parasitize humans can be grouped into four species complexes – *L. tropica*, *L. donovani*, *L. mexicana* and *L. brasiliensis*. Each can give rise to a variety of diseases. Among the best known are oriental sore, an Old World cutaneous lesion caused by *L. tropica* and *L. major*, kala azar, a predominantly visceral form caused by *L. donovani*, and espundia, a New World mucocutaneous infection caused by *L. brasiliensis*. The pathological manifestations of infection show wide variation, analogous in many respects to the spectrum of disease states found in leprosy and similarly associated with the immunological status of the patient. This complexity is compounded by the number of sub-specific forms of *Leishmania*.

The life cycles of leishmanias follow a consistent pattern, all using blood-feeding sandflies, species of phlebotomines, as vectors (Fig. 4.14). In most cases infections are transmitted to humans from a variety of animal reservoir hosts, but direct human–human transmission also occurs. Infections start when flies carrying infective metacyclic promastigotes inject these as they feed. The promastigotes enter macrophages, transform into amastigotes (Fig. 4.15) and commence repeated binary division. When the infected cells eventu-

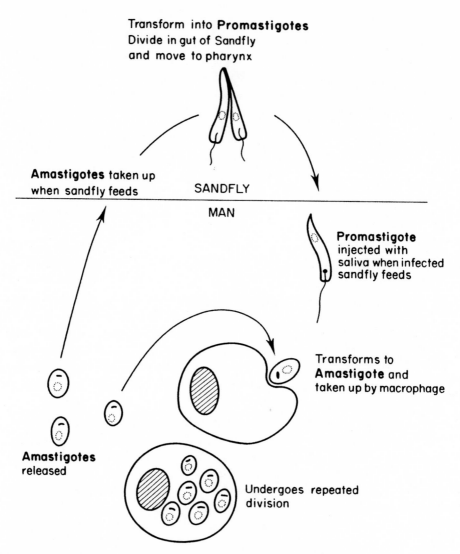

Fig. 4.14 Life cycle of *Leishmania* in sandfly and human hosts.

ally rupture, amastigotes are released and invade other macrophages. In cutaneous and mucocutaneous leishmaniases infection remains localized within the skin and superficial tissues; in *L. donovani* infection spreads to invade visceral organs. Flies acquire infection when they take up infected cells as they feed. The amastigotes are released in the insect's intestine, transform into promastigotes, divide and then move anteriorly into the proboscis.

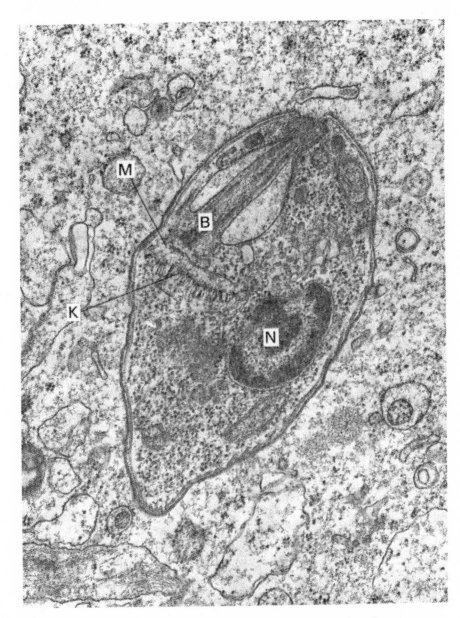

Fig. 4.15   Amastigote of *Leishmania donovani*. B, basal body of flagellum; K, kinetoplast; M, mitochondrion; N, nucleus. (Photograph by courtesy of Professor Dr H. Melhorn.)

L. tropica and *L. major* infections typically give rise to ulcerative sores on the skin at the site of the sandfly bite. Initially there is an infiltration of phagocytes in which the amastigotes multiply, at a later stage lymphocytes accumulate and eventually the lesion resolves and heals. Infections with *L. brasiliensis* can metastasize to the mucous membranes of the nose and mouth, eroding the cartilage. In visceral leishmaniasis parasites spread from the site of initial infection and can become widely distributed throughout the body. Parasite reproduction and cellular infiltration lead to the characteristic enlargement of spleen and liver. A variety of other pathological changes may occur, including anaemia, haemorrhage, intestinal ulceration and damage to the heart.

4.9   Immunity to leishmaniasis

It is difficult to make generalized statements about immunity in the human leishmaniases. Of the species causing cutaneous ulcers *L. major* gives rise to localized lesions that resolve spontaneously and leave the host with a strong immunity to reinfection, but other species (e.g. *L. aethiopica*) can be associated with conditions (e.g. diffuse cutaneous leishmaniasis – DCL) where lesions spread and healing does not occur. Infections with *L. donovani* are progressive and usually fatal if not treated; after cure there is immunity to reinfection. In some cases, however, treatment is followed by the appearance of dermal nodules containing parasitized cells (post-kala azar dermal leishmaniasis). The mucocutaneous leishmaniases, typified by *L. brasiliensis*, are often progressive and non-healing, but recovery can occur with strong immunity to reinfection. In almost all cases immunity shows a rigid species and sub-species specificity.

Infection elicits strong cellular responses, measurable by delayed-type hypersensitivity (DTH) reactions and lymphocyte proliferative responses to leishmanial antigens, but a rather variable antibody response. There is a correlation between cell-mediated responses and ability to develop immunity. Such responses are reduced or absent in non-healing forms of disease (the 'anergic' leishmaniases) where parasite loads are usually heavy, but are well-developed in forms where healing occurs. Parasite-specific antibodies are produced, but these seem to play little or no role in host protection, indeed visceral leishmaniasis is characterized by a marked IgM and IgG hyper-gammaglobulinaemia, which probably reflects a polyclonal activation of B cells.

It is clear that the key to understanding the immunobiology of leishmanial infections lies in the relationship of the parasites to the macrophages in which they live and to the T lymphocytes that regulate macrophage activity. The par-

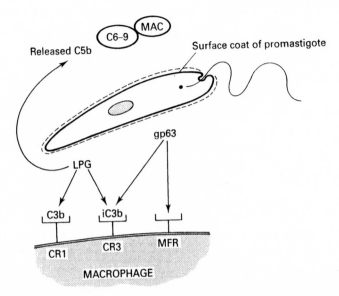

Fig. 4.16 Interactions between *Leishmania* and macrophages. CR1/3, MFR=complement C3b and mannose/fucose receptors; C5b-C9=complement components; MAC=membrane attack complex; LPG/gp63=*Leishmania* surface molecules. (Adapted from Blackwell, 1993, in: *Clinical Aspects in Immunology*, 5th. edn, eds. Lachman *et al.*, Blackwell Scientific Publications, Oxford.)

asite has successfully to enter these professional phagocytes and must then survive in the face of their innate defence mechanisms. We now know that entry occurs after molecules on the surface of the parasite have bound to cell-surface receptors on the macrophage and initiated active phagocytosis. A variety of ligand–receptor combinations can be used, including a number of sugar–lectin interactions, but *in vivo*, the most important involve components of complement (C3b, iC3b) deposited on the parasite surface and complement receptors on the host cell (Fig. 4.16). The complement cascade is activated but, in the case of infective 'metacyclic' promastigotes, does not result in parasite lysis, because long and complex molecules on the parasite's surface coat (lipophosphoglycan – LPG) prevent access of the membrane attack complex to the plasma membrane. These molecules are present in earlier (replicating) promastigotes, but are too short to keep the attack complex away from the membrane and so the parasites are lysed. Amastigotes of *L. donovani* are also complement resistant, but those of *L. major* are sensitive to lysis, which may help to explain the ability of the one to spread to visceral organs and the restriction of the other to local dermal lesions. The use of complement receptors for entry also helps to prevent triggering the respiratory burst and release of toxic oxygen metabolites that normally accompanies phagocytosis. Complement therefore both facilitates parasite uptake by host cells and enables *Leishmania* to avoid a major initial defence mechanism. LPG and other molecules also contribute to inhibition of the oxidative burst.

When completely internalized the parasite lies in a vacuole – the parasitophorous vacuole – and is surrounded, more or less tightly, by host membrane (Fig. 4.17). Lysosomal fusion occurs normally, but the amastigotes survive perfectly well in the phagolysosome, despite the presence of powerful lysosomal enzymes capable of digesting any other particle present simultaneously. (In this respect the survival strategy of *Leishmania* differs from those of *Toxoplasma*, which prevents lysosomal fusion from taking place, and *Trypanosoma cruzi*, which escapes from the vacuole into the cytoplasm). The *Leishmania* parasite seems to have several ways of avoiding damage. LPG is resistant to enzyme activity, and both LPG and the major glycoprotein gp63 (itself a protease) interfere with host cell enzyme function.

Although *Leishmania* parasites are well adapted for survival in macrophages, they can be killed if the host cell is activated. Activation can be brought about in several ways, but during infection the most important is by the action of cytokines from T cells, principally IFN-γ. An understanding of how infections are controlled in this way, and why they may sometimes not be, has helped to explain the diversity of disease states found in human leishmaniasis. The data that have provided this understanding have come very largely from experimental studies with *L. donovani* and *L. major* in the mouse model, in which infections can be established directly by injection of both pro-mastigotes and amastigotes.

4.9.1 *L. donovani*

Visceral leishmaniasis is not fatal in mice, but severe and prolonged infections may develop, the course of infection in a given strain of mouse being genetically determined. Early studies, using a panel of some 20 strains, showed three distinct patterns (Fig. 4.18). In resistant (R) mice there was very little parasite replication in the liver after infection; in susceptible (S) mice replication was unchecked and the animals became massively infected. The difference between the most resistant and most susceptible strains led to a 1000-fold difference in liver parasite load within one month. S strains then diverged, some developing resistance (SR) and controlling the infection, others remaining completely susceptible (SS) and maintaining a long-term heavy parasite burden.

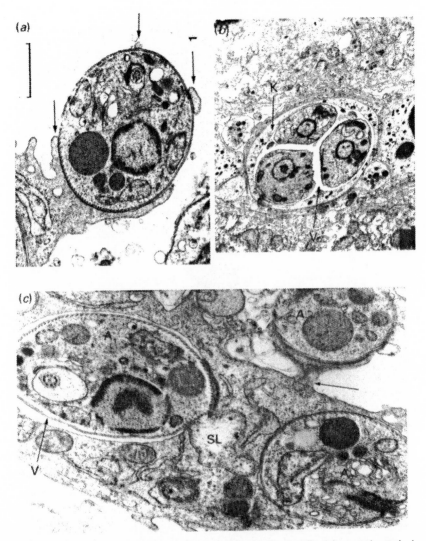

Fig. 4.17   Relationship of *Leishmania* amastigotes to host cells. (a) amastigote being taken up by macrophage pseudopodia (arrowed); (b) amastigotes lying within a parasitophorous vacuole (V) within a Kupffer cell (K); (c) two amastigotes (A_1, A_2) within a macrophage 30 minutes after *in vitro* infection and a third (A_3) being taken up by pseudopodia (arrowed). The intracellular parasites lie within membrane-bound parasitophorous vacuoles (V). A secondary lysosome (SL) is seen fusing with the vacuole and releasing its contents. (Photograph (a) from Bray & Alexander, 1983, in *Leishmaniasis*, eds. Killick-Kendrick & Peters, Academic Press, New York, by permission of the authors and publishers. Photograph (b) by courtesy of Dr S. Croft. Photograph (c) from Blackwell & Alexander, 1983, *Transactions of the Royal Society of Tropical Medicine and Hygiene*, **77**, 636 by permission of the authors and publishers.)

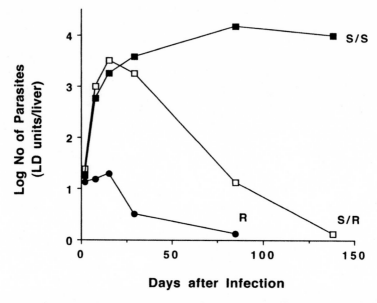

Fig. 4.18   Schematic representation of course of infection with *Leishmania donovani* in strains of mice showing genetically determined susceptibility (S/S, S/R) or resistance (R). S/S mice have little or no innate resistance and do not acquire immunity, remaining susceptible and developing a chronic infection. S/R mice, in contrast, do develop immunity and eventually clear the infection. R mice show a strong innate resistance.

The initial distinction between (R) and (S) strains appears well before the intervention of immunity and reflects an innate difference in the capacity of macrophages to support parasite development. In the first two weeks, replication in R mice is less than 8-fold, in S mice it is 80- to 100-fold. This difference is determined by a single gene, originally designated *Lsh*, with two alleles, *Lsh*[r] (incompletely dominant and giving resistance) and *Lsh*[s]. *Lsh* has been mapped to chromosome 1, is expressed in liver Kupffer cells, and has been shown to be identical with the genes designated *Ity* and *Bcg*, which regulate resistance of mice to *Salmonella typhimurium* and *Mycobacterium* spp. respectively. *Lsh*, *Ity* and *Bcg* are now identified with the gene *Nramp* (natural resistance associated macrophage protein). This codes for a membrane protein and appears to operate through an influence on the degree to which cells respond to receptor-mediated priming, and thus regulates expression of non-specific mechanisms of resistance involving oxygen radicals and nitric oxide. Following infection there is a time-lag of about two days before differences in parasite development in R and S cells becomes apparent. This presumably reflects the

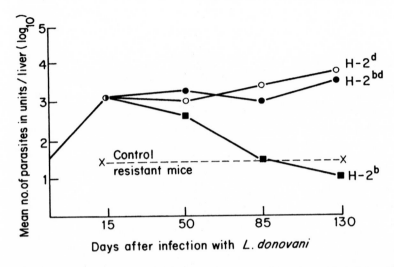

Fig. 4.19   MHC (H-2)-linked inheritance of immunity to infection with *Leishmania donovani* in mice. Course of infection in H-2^d, H-2^b and hybrid H-2^{bd} mice of the C57 BL/10 ScSn genetic background. (Data from Blackwell *et al.*, 1980, *Nature*, **283**, 72.)

time necessary for signal transduction, gene expression, response, and effect on the parasite. *Lsh* also regulates expression of MHC Class II molecules, and thus exerts a marked influence upon the antigen presentation function of infected macrophages, *Lsh'* cells being much more effective.

The difference between S strains in their ability to control infection after the initial phase of replication is dependent upon T helper cell-mediated responses, and is influenced by particular MHC-linked genes. In mice of the B10 genetic background, for example, those expressing the H-2^b haplotype control infections rapidly whereas mice with the H-2^d haplotype maintain heavy infections for more than 4 months. H-2^{db} heterozygotes behave like H-2^d mice, i.e. susceptibility is dominant (Fig. 4.19). As with *L. major* (see below) this genetic control may operate via selective induction of either host-protective or disease-promoting Th subsets. Host protection would therefore reflect Th1 subset activity and the production of macrophage-activating IFN-γ, stimulation of Th2 cells would lead to IL-4 production, down-regulation of IFN release and failure to cure infection. How such a subset bias comes about is still unknown, but it may arise from interference with normal antigen presentation. Infected macrophages express less MHC Class II on their surface, are less responsive to IFN-γ than normal cells, and so are likely to become overloaded with parasite antigen.

4.9.2   *L. major*

As with *L. donovani*, mouse strains show clear-cut genetically determined variation in response to infection. In some highly-resistant strains infections do not
establish at all; in resistant strains there is development of the characteristic
skin lesion, which then resolves; in susceptible strains (notably BALB/c) the
lesion does not heal, visceral metastasis occurs, and death results. The genetic
control of resistance to *L. major* is quite distinct from that of *L. donovani* and
strains of mice may show quite opposite phenotypes when infected with the
two species. As with *L. donovani* there is separate control of innate resistance
and acquired immunity, but MHC-linked genes have much less influence. The
major gene controlling healing/non-healing is termed *Scl-1*, and it has recently
been mapped to chromosome 11. Like *Lsh, Scl-1* regulates the response of
macrophages, but it may have wider pleiotropic effects and influence infection
at several points. The gene has a major influence on the growth of the lesion
during the early stages. Macrophages from susceptible strains allow far greater
parasite replication *in vitro* than cells from resistant strains (Fig. 4.20). *In vivo*
this effect is compounded, possibly by an effect upon release of chemo-
attractants for immature cells of the macrophage/monocyte series which can
readily be infected but respond poorly to cytokine stimulation, allowing the
parasite population to expand freely.

   As with *L. donovani*, immunity to *L. major* is T cell-mediated. Both CD4$^+$
and CD8$^+$ cells contribute to this immunity, but the former have the most
important role. Athymic nude mice are highly susceptible to *L. major*, but prior
transfer of very small numbers ($< 3 \times 10^5$) of normal CD4$^+$ cells provides
effective protection. Similarly, CD4$^+$ cells from immune mice transfer protection adoptively into naive recipients. Depletion of CD4$^+$ T cells *in vivo* by treatment with anti-CD4 monoclonal antibodies removes the ability of mice to
control lesion development. Strains of immunologically normal mice show
considerable genetically determined variation in ability to control infection
and it is now known that this reflects the nature of the CD4$^+$ T cell response
elicited by exposure to the parasite. Resistant strains, such as C57/BL6 or
CBA, produce resistance-promoting Th1 cells, which release IFN-γ, allowing
normal macrophage activation and parasite killing to take place. The Th
response of susceptible BALB/c mice is biassed to the Th2 subset. These
cells are disease-promoting, largely because the cytokines produced both shut
off the potential for IFN-γ release (IL-4) and promote production of immature bone-marrow derived macrophages (IL-3 and GM-CSF) in which the
parasite can replicate. Elegant experiments have shown that the response

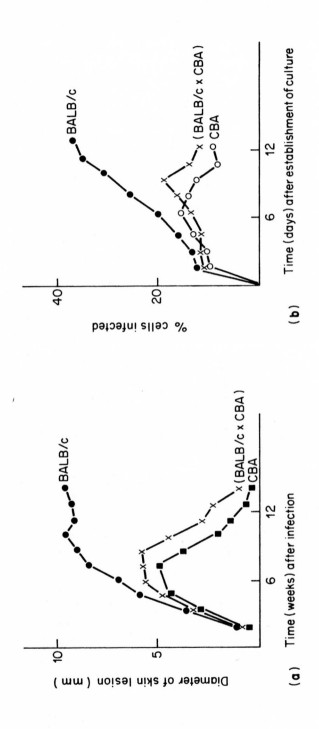

Fig. 4.20 Genetic control of immunity to *Leishmania major* in mice.
(a) Growth of skin lesion in infected BALB/c, CBA and (BALB/c x CBA) F₁ mice. (b) Time course of *in vitro* infection of skin macrophages taken from the same strains of mice (Redrawn from Gorczynski & MacRae, 1982, *Cellular Immunology*, **67**, 74.)

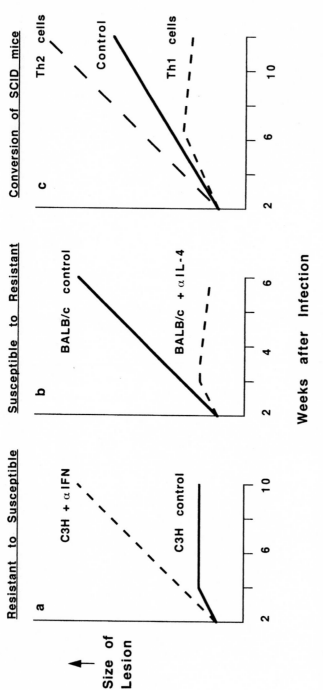

Fig. 4.21 T helper cell subset and cytokine influences on susceptibility and resistance to *Leishmania major* in mice. (a) Treatment of genetically resistant C3H mice with anti-interferon gamma (IFN-γ) antibody results in susceptibility to infection and progressive lesion growth. (Data from Scott, 1991, *Journal of Immunology*, **147**, 3149.) (b) Treatment of genetically susceptible BALB/c mice with anti-interleukin-4 (IL-4) antibody results in resistance and healing of lesion. (Data from Chatelain et al., 1992, *Journal of Immunology*, **148**, 1182.) (c) Reconstitution of immune-deficient SCID mice with cloned T cells from BALB/c mice results in resistance and healing of lesion if cells with the characteristics of T helper subset 1 (Th1) are used, but increases susceptibility and allows greater lesion growth if Th2-type cells are used. (Data from Holaday et al.,1991, *Journal of Immunology*, **147**, 11653.)

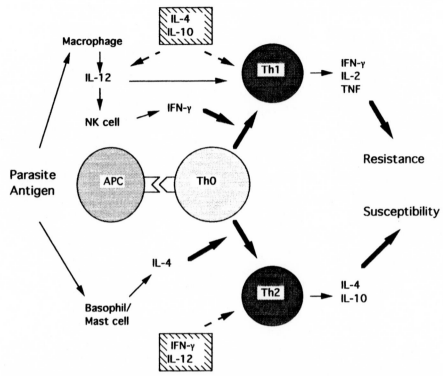

Fig. 4.22 Cytokine-mediated interactions between parasites, non-lymphoid cells and T cell subsets in determining the outcome of infection with *Leishmania* in mice. →, positive influence on target cell; -→, negative influence on target cell.

phenotypes of mouse strains can be switched between resistance and susceptibility by injecting the appropriate recombinant cytokines or anti-cytokine antibodies. Thus, injection of rIFN or anti-IL-4 antibody allows BALB/c mice to express resistance, injection of IL-4 or anti-IFN induces susceptibility in CBA mice (Fig. 4.21). Many other experiments (e.g. using immunization by different routes, irradiation or cytotoxic drugs to manipulate responsiveness) have shown that susceptible mice have the potential to develop protection-promoting Th cells, but under normal conditions this potential is not realized and disease-promoting Th2 cells are induced instead – why this occurs is still not fully understood. It is likely that quantitative, and possibly qualitative, differences in antigen presentation by heavily infected macrophages may influence subset induction, as may antigen presentation via B cells. Certainly it is now known that polarization to the Th1 response is assisted by release of IFN-γ from NK cells, in response to production of IL-12 from stimulated macrophages, and it is possible that polarization to a Th2 response may follow a corresponding release of IL-4 from non-lymphoid cells such as basophils and mast cells (Fig. 4.22).

The relevance of murine studies to the human leishmaniases is supported by many observations showing individual or racial differences in response to infection, by evidence that homologues of the resistance genes identified in mice exist in the human genome, and by correlations between the cytokine responses to *Leishmania* infections seen in humans and in mice. Studies with New World cutaneous leishmaniasis have shown that cytokine responses of patients with the diffuse form of the disease (DCL) are biassed to Th2 cytokines, whereas those of patients with localized lesions show a Th1 bias. Work with mouse models has therefore provided a valuable conceptual framework for work that may lead directly to improved approaches to management of leishmanial diseases.

5

African trypanosomes
Antigenic variation

5.1 Introduction

Although the parasites discussed in the previous chapter are primarily intracellular organisms, they necessarily spend short periods outside the host cell. At these times they are demonstrably vulnerable to immunological attack. The African trypanosomes (genus *Trypanosoma*, section Salivaria) are the most important protozoa that live as extracellular parasites throughout the entire life cycle. Their long-term survival in this potentially hostile environment is primarily attributable to their ability to undergo antigenic variation, and thus keep one step ahead of the host's immune response. Our knowledge of this phenomenon represents, to date, one of the most completely understood of all parasite survival strategies.

Trypanosomes are flagellate protozoa, with a structural organization that is comparatively simple, even when examined under the electron microscope. The life cycle typically includes a phase spent in the bloodstream and tissue fluids of a mammalian host, during which there is multiplication by binary fission, and a phase spent in the body of a vector arthropod (Fig. 5.1). For the majority of clinically and economically important species the arthropod concerned is a tsetse fly and the parasite undergoes major morphological and biochemical changes in this host, as well as repeated division. The parasites are taken up by the fly when it feeds on the blood of the mammalian host. Bloodstream forms (trypomastigotes) are characteristically long and slender (Fig. 5.2), with the kinetoplast situated posterior to the nucleus. The long, tubular mitochondrion is non-functional, the citric acid cycle and cytochrome system are absent and respiration occurs by glycolysis. A proportion of trypo-

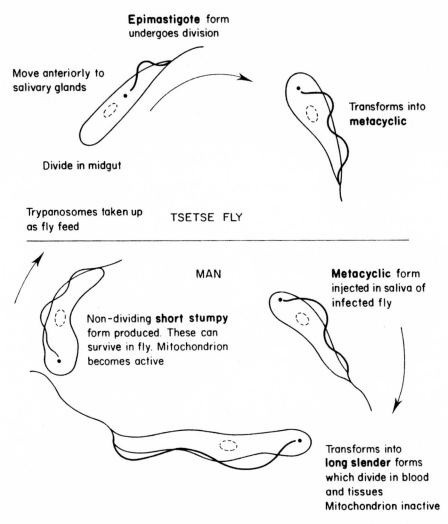

Epimastigote form
undergoes division

Move anteriorly to
salivary glands

Transforms into
metacyclic

Divide in midgut

Trypanosomes taken up
as fly feed

TSETSE FLY

MAN

Non-dividing **short stumpy**
form produced. These can
survive in fly. Mitochondrion
becomes active

Metacyclic form
injected in saliva of
infected fly

Transforms into
long slender forms
which divide in blood
and tissues
Mitochondrion inactive

Fig. 5.1   Life cycle of *Trypanosoma* in tsetse fly and man.

mastigotes are shorter, stumpy forms, and in these there is partial redevelopment of mitochondrial function. Only stumpy forms survive uptake by the tsetse fly. In the midgut the mitochondrion is functional and there is reappearance of the Krebs cycle and the cytochrome electron transport system. After extensive multiplication the midgut forms migrate anteriorly, between the peritrophic membrane and the gut wall and enter the salivary glands. There they differentiate, first into the epimastigote (anterior kinetoplast) and then, after further division, into the infective metacyclic form. Transmission is

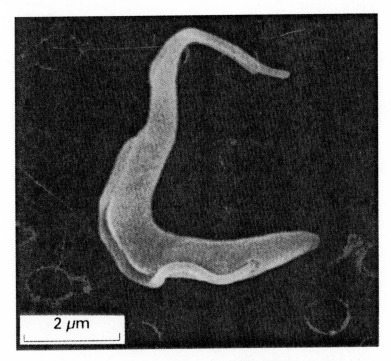

Fig. 5.2   Scanning electron micrograph of bloodstream form of *Trypanosoma brucei brucei*. (Photograph by courtesy of Dr L. Tetley and Professor K. Vickerman.)

achieved when the infected tsetse next feeds, the metacyclics being introduced directly into the host with the saliva. (Experimental infections can be initiated directly by syringe passage of bloodstream forms.)

Tsetse flies occur only in a zone of equatorial Africa – the tsetse belt – and the requirement for cyclical development in this host restricts most trypanosome infections to this continent. However, direct transmission via the mouthparts of other biting flies is possible. In some species the requirement for cyclical development has been lost; transmission is therefore purely mechanical and can be achieved by a variety of blood-feeding flies and even by vampire bats. Such species, e.g. *T. evansi* and *T. equinum*, can survive outside the tsetse belt. In *T. equiperdum* transmission is achieved without the intervention of a vector and occurs venereally in equine hosts.

5.2   Trypanosomiasis

African trypanosomes are widespread in game animals, where they appear not to be unduly pathogenic. Trypanosomiasis as a recognizable and serious disease occurs in man, where it is caused by *T. brucei rhodesiense* and *T.b. gambiense*, and in domestic stock, where it is caused by several species, including *T.b. brucei*, *T. congolense* and *T. vivax*. The only major differences between the sub-species of *T. brucei* appear to be infectivity for man and the nature of the disease caused by the two human parasites. Infections are associated with fever, anaemia, inflammation of lymph nodes, loss of weight and the lethargic condition from which the popular name of sleeping sickness is derived. Although trypanosomes appear initially in the blood they also move into the extravascular spaces of deeper tissues and frequently invade the central nervous system. It has been suggested that these tissue infections form a reservoir of dividing stages from which the blood infection is maintained.

Untreated infections are normally fatal, the time to death varying considerably. In man *T.b. gambiense* gives rise to a chronic infection, whereas infections with *T.b. rhodesiense* follow an acute course. There is no doubt that trypanosomiasis has a profound impact upon human development in Africa, not only through its direct effects in terms of human disease, but also indirectly through bovine infections, which prevent the utilization of grasslands for cattle farming. An estimated 35 million people and 25 million cattle are at risk from infection, and some 3 million cattle die annually from the disease.

5.3   Antigenic variation

A characteristic feature of chronic trypanosome infections in humans and animals is the occurrence of regular fluctuations in the level of parasitaemia (numbers of parasites present in the blood – Fig. 5.3). As long ago as 1907 it was suggested that each fall in parasite numbers was brought about by the action of host antibodies, the subsequent rise being due to proliferation of parasites that differed antigenically from their predecessors. Subsequent research has shown this to be a remarkably accurate interpretation, and sequential antigenic variation is now known to be a feature of all salivarian trypanosomes. Variation does not appear to occur in the stercorarian trypanosomes such as *T. cruzi*, which employ other strategies for long-term survival.

Antigenic variation can be detected directly by immunological techniques.

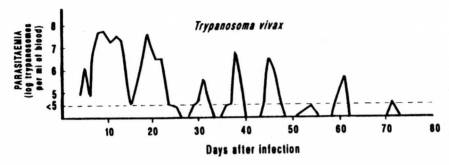

Fig. 5.3   Course of parasitaemia in a cow infected with *Trypanosoma vivax*, showing repeated peaks as antigenically different populations of trypanosomes appear, grow and are eliminated by immunity (Redrawn from Barry, 1986, *Parasitology*, **92**, 51).

Binding of antibody specific for a particular variant antigenic type (VAT) can be measured *in vitro*, by immunofluorescence, agglutination, neutralization of infectivity or complement-mediated lysis. Trypanosomes carrying a different VAT, e.g. populations taken at different time points after infection, show no antibody binding, thus validating the interpretation of fluctuating para-sitaemias described above. *In vivo*, trypanosomes binding antibody would be removed from the bloodstream, those unaffected would multiply to provide the next peak, before in turn being removed by antibodies specific for their particular VAT.

5.3.1   The surface glycoprotein coat

Binding of antibodies to intact bloodstream trypanosomes implies the exis-tence of antigenic material on the surface of the parasite. This can be visual-ized directly by EM techniques using labelled antisera, which reveal binding to a dense surface coat external to the cell membrane of the parasite (Fig. 5.4). This coat is formed by a closely packed monolayer of over 10^7 molecules, all belonging to a single species of glycoprotein that has an MW of 55–60 kDa. It is 12–15 nm thick and makes up about 10% of the dry weight of the organism. Trypanosomes therefore have about 10 times as much surface glycoprotein as the RBC (red blood cells), which have a comparable surface area, and it is obvious that this abundance must reflect some major biological function. In essence this is to provide a macromolecular barrier, preventing host molecules from interacting with the parasite's plasma membrane. We now know that the glycoprotein coat is essential for life in the host's bloodstream, both because it prevents complement-mediated lysis and is antiphagocytic, and because it

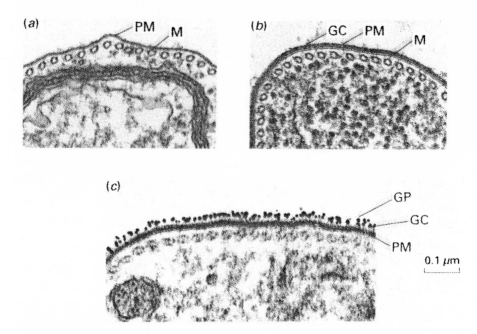

Fig. 5.4 Transmission electron micrographs of the surface coat of *Trypanosoma brucei brucei*. (a) 26°C culture form in which the glycoprotein coat has been lost. (b) Bloodstream form showing the thick glycoprotein coat covering the plasma membrane. (c) Immunogold labelling of bloodstream form (5 nm gold particles conjugated to a monoclonal anti-VSG epitope antibody). The glycoprotein coat is visible beneath the particles. GC, glycoprotein coat; GP, gold particle; M, microtubule; PM, plasma membrane. (Photographs by courtesy of Dr L. Tetley and Professor. K. Vickerman)

provides the molecular basis for antigenic variation, allowing the parasite population to survive in the face of immune recognition and attack.

Trypanosomes belonging to a particular VAT express only a single variant specific glycoprotein (VSG), and epitopes on this VSG are the target of VAT-specific antibodies. Antigenic variation results in the expression of a different VSG in the surface coat of some trypanosomes, thus generating a different VAT that is unaffected by antibody against the previous VAT. Thus, even though each VSG is the focus for destructive immune responses, injection of as little as 3 μg of purified VSG conferring complete protection on mice exposed to the homologous VAT, these responses are highly specific and ineffective against parasites of a different VAT. It seems obvious that, in transmission terms, it is better for the parasite to persist for long periods with a constantly fluctuating parasitaemia, than either to replicate in an uncontrolled

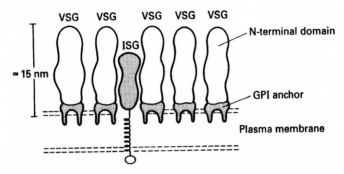

Fig. 5.5 Organization of the variant specific glycoprotein (VSG) molecules in the surface coat of a trypanosome. Each VSG molecule is anchored by a glycosylphosphatidylinositol (GPI) residue. For every 100 VSG molecules there is a single invariant surface glycoprotein (ISG) that has a transmembrane helix and a short intracellular domain. (Redrawn from Overath *et al.*, 1994, *Parasitology Today*, **10**, 53.)

way, as do virulent strains, and kill the host, or maintain low-level infections that may be eliminated by immunity.

The stucture of the VSG is now known in great detail. It consists of a polypeptide chain of about 500 amino acids, divided into N-terminal and C-terminal domains, anchored in the plasma membrane by a glycosylphosphatidylinositol (GPI) glycolipid anchor. The N-terminal domain is a highly variable region and there is very little homology in amino acid sequence between different VSGs. In contrast, the C-terminal domain is a constant region, showing considerable sequence homology between VSGs. Both the polypeptide backbone and the GPI anchor carry carbohydrate side-chains. X-ray crystallographic studies have shown that VSG molecules exist on the cell suface as cylindrical, dimeric structures, orientated at 90° to the plasma membrane (Fig. 5.5).

The close packing of the molecules means that only part of the N-terminal domain of each is exposed, effectively providing the trypanosome with a continuous, proteinaceous external surface. The epitopes recognized by VAT-specific antibodies are therefore those expressed on these exposed regions of the VSG. The variability of amino acid sequence in these regions accounts for the separate antigenic specificity of each VSG. Cross-reacting epitopes do occur on VSG molecules. These are primarily associated with the carbohydrate side-chains, particularly those of the GPI anchor. They are not exposed in the intact trypanosome, and accordingly are not available for interaction with host antibodies, even though specific antibodies to these cross-reacting determinants, raised to shed VSG molecules, do arise during infection.

The VSG coat is maintained throughout the bloodstream phase, but is shed when trypanosomes are taken up by the insect vector (as it is in culture – Fig. 5.4a). During the early (procyclic) stages of development in the tsetse the coat is gradually replaced by molecules of another glycoprotein – procyclin. Later on, as the parasite changes into the metacyclic stage, procyclin is lost and the VSG coat resynthesized. Stages without VSG are rapidly lysed by complement if brought into contact with blood.

5.3.2   Expression of VSGs during infection

One of the major questions concerning antigenic variation has been the nature of the mechanisms, in both host and parasite, responsible for the sequential appearance of VSGs during the course of infection. Progress in answering this question has come from the application of immunological and molecular techniques to analyse host response and parasite biology.

An early suggestion was that antigenic variation was induced in a population by the appearance of antibodies against the surface glycoproteins. The bulk of experimental evidence does not support this idea. Variation is seen in populations established in irradiated (and therefore immunologically incompetent) hosts as well as in trypanosomes passaged between hosts at intervals that are shorter than that required for the expression of antibody responses. Antigenic variation has also been recorded in bloodstream forms maintained wholly *in vitro*. It therefore appears that the major effect of antibody *in vivo* is to eliminate trypanosomes carrying the predominant VSG, allowing multiplication of parasites expressing different antigens. Detailed studies on infections initiated with cloned parasites, using sensitive serological techniques to identify specific VSGs, have shown that although the population present during the rising phase of a parasitaemia peak consists predominantly of a single VAT (the homotype), minor VATs (heterotypes) are also present. One of these heterotypes will become the next homotype when the population crashes, the relative growth rates of those present determining which becomes dominant (Fig. 5.6).

Antigenic variation occurs in infections established with rodent-adapted cloned parasites at a frequency (about 1 in 10^6 organisms) that could be generated by genetic mutation alone. However, many lines of evidence suggest that this is not the case, not least that frequency values are much higher in more recently isolated populations. One of the strongest arguments against mutation as the major cause of variation comes from observations that, with defined infections, certain VATs are expressed consistently during the early

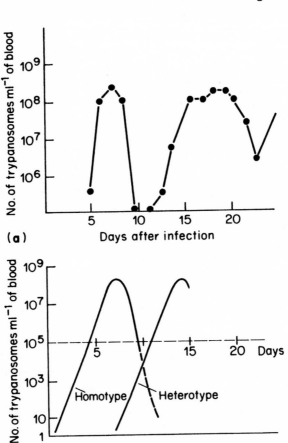

Fig. 5.6   (a) Course of parasitaemia in mouse infected with a single trypanosome on day 0. First peak is distinct, due to growth of predominant VAT (homotype). Relapse peaks less distinct, because of overlapping growth of heterotypes. (b) Theoretical picture showing growth of homotype and relapse arising from growth of heterotype. Only one heterotype is shown, in practice several might be present simultaneously. One heterotype becomes the predominant VAT of the relapse peak that follows remission of the original homotype. 10^5 represents the detectable level of patent parasitaemias. (Both redrawn from Vickerman, 1978, *Nature*, **273**, 613.)

stages and are then followed by a regular and reasonably predictable sequence of predominant VATs, at least until the later stages of a chronic infection. When particular VATs are cloned, and one clone used to establish infections in different individual hosts, the sequence of VATs produced in each host is similar. Each clone seems to be able to express a characteristic repertoire of

Table 5.1 *Appearance of selected VATs in bloodstream* T. b. brucei *following infection of mice by tsetse flies fed on donors with a VAT-defined population*

Days after infection	AnTat 1 VATs (% of trypanosomes showing positive fluorescence with mono-specific antisera)						
	1.2	1.6	1.10	1.14	1.20	1.30	1.45
3	0	0	0	0	0	28	10
4	0	1	0	9	0	28	6
6	1	3	0	28	0	0	0
8	7	9	0	0	7	0	0
10	14	10	0	0	0	0	0

Notes:
VAT homotype ingested by fly: – AnTat 1.14.
VATs detected in metacyclics from fly: 1.6 (6%), 1.14 (0%), 1.30 (23%), 1.45 (5%).
Source: (Data from Hajduk & Vickerman, 1981, *Parasitology* **88**, 609.)

VATs (collectively referred to as a serodeme), and this repertoire is presumably determined by its genotype. The total number of VATs actually expressed may be very large, for example more than 100 have been recorded in infection with a cloned line of *T. equiperdum*, and the potential VAT repertoire in a species may be enormous.

A further argument against mutation is the fact that the VATs expressed by metacyclic trypanosomes represent only a small subset of the total repertoire found in the bloodstream forms. In one series of experiments with a cloned population of *T. b. rhodeiense*, the maximum total metacyclic repertoire was estimated as being 27, based on reactivity with a panel of monoclonal antibodies. However, 95% of the metacyclics tested were labelled by only ten of these antibodies, indicating that the expressed repertoire is dominated by only a few types. After infection new VATs appear rapidly, one of the first to do so being that originally ingested by the transmitting fly (Table 5.1). It is now firmly established that antigenic variation is achieved by a complex mechanism of gene switching, not by mutation, and that it is independent of antibody induction.

5.3.3   Genetic control of antigenic variation

In the trypanosome genome, which is diploid, there are separate genes (or small gene families) for each VSG, the total potential repertoire of genes numbering several hundreds. These are carried on the 20 or so large chromosomes as well as on the numerous minichromosomes. Typically only one gene is expressed at any given time, the other genes in the VSG repertoire (the basic copy genes) occurring in tandem arrays within the chromosomes where they are not expressed in this position. In order for one of these basic copy genes to be expressed it has to be copied (as an expression-linked copy – ELC) into a transcriptionally-active specific expression site located at the chromosomal telomere – the process known as gene conversion (Fig. 5.7a,b). When this occurs, the existing ELC in this site is lost or inactivated, and production of a new VSG is initiated by transcription of the replacement ELC. Other ELC genes in telomeric but 'silent' expression sites can be transcribed if their site becomes active, the previously active site being turned off (Fig. 5.7c). These telomeric genes are, by virtue of their position, likely to be involved in recombinational events; switching of genes into an active expression site by this means is another way in which antigenic variation is initiated (Fig. 5.7d).

In a given serodeme, therefore, all of the trypanosomes may carry all of the genes necessary for the VSG repertoire characteristic of that serodeme. In each trypanosome, however, only one gene will be functionally active, and only one VSG mRNA detectable. This selective expression is regulated at the transcriptional level. At any one time during infection, the gene expressed and the mRNA present will be the same in the vast majority of organisms. Switching of genes is not random, expression of new VSG genes and the appearance of new VATs occurring with some degree of predictability. Which of the new VATs comes to dominate the population, when the previous VAT is destroyed, will be determined by the switching hierarchy and by differences in switching rates between pairs of VATs. Long-term survival of trypanosomes therefore involves a delicate balance between the rate of gene switching and the rate at which the host responds immunologically.

5.4   Mechanisms of protective immunity

The failure of hosts to control trypanosome infections results primarily from the phenomenon of antigenic variation and not from an inability to mount protective responses. The effectiveness of such responses can be seen under

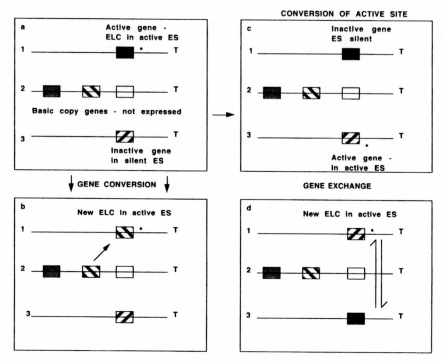

Fig. 5.7 Antigenic variation in trypanosomes. Three ways in which new genes coding for VSGs can be expressed are shown in a diagrammatic representation of four VSG genes on three chromosomes (ELC, expression linked copy gene; ES, expression site; T, telomere): (a) active ELC transcribed in active ES; (b) gene conversion – existing ELC in active ES replaced by a basic copy gene; (c) conversion of active site – ES activated on different chromosome; (d) gene exchange – reciprocal exchange of genes between active and silent ES.

controlled experimental conditions when immunity is stimulated by injection of purified VSG or by vaccination with irradiated trypanosomes, or when antibodies are transferred. In all of these cases strong protection is given, but only against the VAT concerned.

Laboratory studies in rodents show that immunity to trypanosomes is entirely antibody-dependent. Appearance of VAT-specific antibody is temporally associated with disappearance of that VAT from the bloodstream; passive transfer of antibody, or adoptive transfer of B cells (but not T cells), restores immunity to immunodeficient recipients. The ability of antibody to clear trypanosomes from the blood is striking. For example, injection of a VSG-specific monoclonal antibody into mice infected with a moderate parasitaemia of *T. brucei brucei* (10^7 parasites ml^{-1}) resulted in clearance within 20 minutes. IgM antibodies are the most effective, and appearance of this isotype during infection is rapidly followed by parasite removal; IgG antibodies appear after this elimination has occurred. The high concentrations and prolonged production of IgM during infection probably reflect a T-independent stimulation of B cells brought about by presentation of the serially repeated epitopes exposed at the surface of the glycoprotein coat. The ability of thymus-deficient nude mice to control an initial parasitaemia, which they can do as successfully as normal mice, and to respond protectively to immunization against VSG reflects this production of T-independent antibody.

It seems most probable that elimination of trypanosomes following antibody binding is achieved primarily through phagocytosis by Kupffer cells in the liver. Direct injection of parasites metabolically labelled with an isotope of methionine into mice that had been passively immunized with specific antibody showed a rapid accumulation of radiolabel only in the liver and not, surprisingly, in the spleen (Fig. 5.8). Electron microscope studies have shown that breakdown of antibody-coated trypanosomes begins almost immediately following their uptake into the phagolysosomes of the Kupffer cells. Phagocytosis does not require complement, although complement-mediated mechanisms may play an additional role, suggesting that IgM Fc receptor binding is the primary mechanism involved in uptake.

Cell-mediated responses appear to play no major role in protective immunity, although they may contribute to the development of the primary chancre, the lesion which appears in natural hosts at the site of the tsetse bite, in which localization and destruction of some parasites may occur.

5.5   Immune modulation by trypanosomes

Trypanosome infections give rise to pronounced depression of immune responses in humans, bovines and mice, affecting responses to a variety of heterologous antigens, including vaccines, other infectious organisms and experimental antigens. Paradoxically, immune depression is coincident with very substantial increases in circulating immunoglobulin, particularly IgM. Indeed, in endemic areas, IgM hypergammaglobulinaemia can be diagnostic of trypanosomiasis.

The production of antibody responses to VSGs and the frequent antigenic variation that occurs in chronic, relapsing infections would themselves be expected to lead to the large increase in circulating immunoglobulin. However, the level of this increase in experimental infections is such that only a proportion of the immunoglobulin is likely to be specific anti-trypanosome

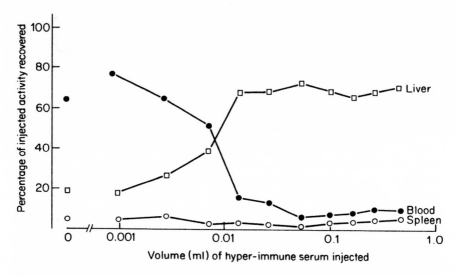

Fig. 5.8 Distribution of radio-labelled (^{75}Se) trypanosomes (*T. b. brucei*) 60 minutes after injection into mice passively immunized with various amounts of serum from rats immunized against the homologous parasite. (Redrawn from MacAskill *et al.*, 1980, *Immunology*, **40**, 629.)

antibody, the remainder being immunoglobulin produced as a result of parasite-induced polyclonal activation of B cells. Experimental studies in mice have shown that, within a week of infection with 10^5 *T. b. brucei*, the spleen contained elevated numbers of IgM-secreting cells specific for heterologous antigens such as sheep RBC, chicken gamma globulin, pneumococcal polysaccharide and the hapten TNP. Autoantibodies against DNA, RBC and T cells were also detected. The levels of polyclonal antibody synthesis were comparable in infected nude and normal mice, implying that the effect of infection was independent of any regulatory T cell activity.

These modulatory influences probably reflect a combination of trypanosome-mediated activities. Trypanosomes exert mitogenic effects on both T and B cells, and also interfere with antigen processing and presentation. Macrophages from infected animals have been shown *in vitro* to exert suppressor activities, and T cell functions decline as infection progresses. All of these effects depend upon the presence of live trypanosomes, and are rapidly reversed after successful chemotherapy.

An important question is the influence that immune depression exerts upon responses to the trypanosomes themselves. In mice infected with some serodemes there is evidence of a progressive decline in ability to control waves of parasitaemia, but this may be a consequence of the relatively high parasite load in this host and the rate at which new antigenic variants arise. Cattle, while showing depression of responses to heterologous antigens, maintain the ability to control infection, and continue to produce both IgM and IgG anti-parasite antibody.

6

Schistosomes

Concomitant immunity and immunopathology

6.1 Introduction

The immunobiology of infections with helminths differs in two important respects from that of infections with protozoa. Helminths are many times larger than protozoans and do not replicate within the vertebrate host. The difference in size restricts the ways in which hosts can mount effective immune responses and the lack of replication influences the survival strategies of both host and parasite. These points will be considered initially in the context of infections with schistosomes.

Schistosomes are platyhelminth (flatworm) parasites that live as adults in the blood vessels of their mammalian and bird hosts. The majority of parasitic flatworms are hermaphrodite, but schistosomes are not. Their generic name, which means 'split-body', describes the appearance presented by the individual males and females, which live as permanent couples, the larger, fatter male carrying the female in a longitudinal groove. The life cycle is indirect and involves a molluscan intermediate host (Fig. 6.1). Eggs are released from the body of the final host via faeces or urine, depending upon the species of schistosome. They are mature when released and, on contact with fresh water, hatch to liberate the first larval stage, the miracidium. This small, ciliated organism has limited powers of survival and must come into contact with the correct species of aquatic snail if the cycle is to continue. When contact is made, the miracidium penetrates into the snail, losing the ciliated epidermis as it does so, and transforms into the sporocyst. The sporocyst has a syncytial, cytoplasmic outer surface and, having no digestive system, absorbs nutrients directly across this layer. Within its body, permanently embryonic cells divide

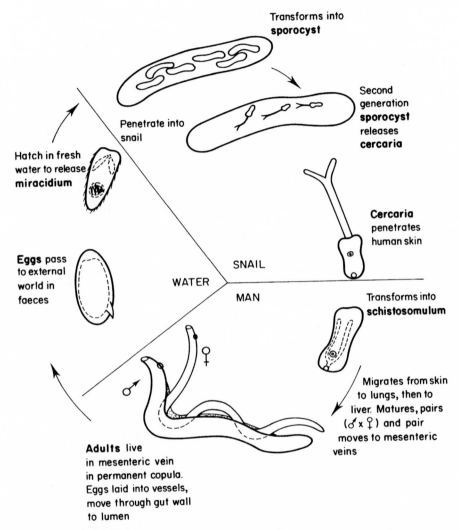

Fig. 6.1 Life cycle of schistosome in snail and man (based on *S. mansoni*).

and differentiate to form a second generation of sporocysts. In these, similar cells divide and differentitate to form the next larval stage, the cercaria. By this process of asexual division the schistosome can greatly increase its reproductive potential. In *S. mansoni*, a species that infects man, it has been calculated that one miracidium is capable of producing 200 000 cercariae. The cercaria is the stage infective to the final host. It leaves the snail and becomes temporarily free-living. The larva shows characteristic behaviour patterns,

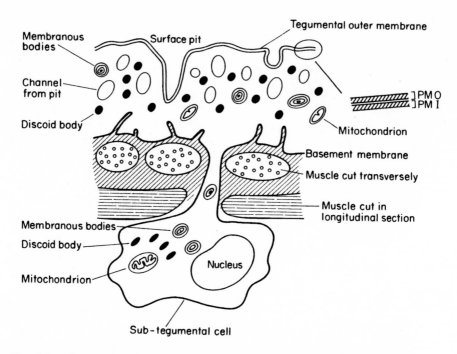

Fig. 6.2 Diagrammatic representation of the tegument of an adult schistosome seen at the ultrastructural level. PMI, PMO, Inner and outer plasma membranes. (Redrawn from McLaren, 1980, *Schistosoma mansoni: The Parasite Surface in Relation to Host Immunity*, Wiley, London.)

hanging from the surface film of the body of water into which it has been released, detaching, sinking, then swimming upwards again. By so doing it may increase the chances of coming into contact with a suitable host.

Infection of the final host occurs by direct penetration of the skin. The cercaria first attaches to the epidermis, sheds its tail and then, by a combination of vigorous movement and enzyme secretion, penetrates through the epidermis, crosses the basement membrane and enters the dermis. On penetration the cercaria undergoes profound changes in the structure and physiological properties of the surface layer, the tegument. Whereas the cercarial surface is bounded by a trilaminate plasma membrane, bearing a thick glycocalyx, the tegument of the succeeding stage, the schistosomulum, develops a multilaminate plasma membrane, the glycocalyx is lost and the larva becomes unable to survive in water (Figs. 6.2 and 6.3). After a period of time the schistosomulum migrates via the blood stream to the lungs and from there to the liver. Following growth and maturation in this organ, the adult worms pair and

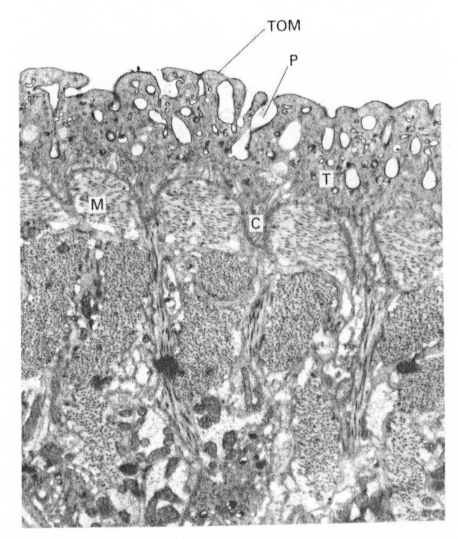

Fig. 6.3   The complex tegument of *Schistosoma mansoni* (female) showing the structures illustrated diagrammatically in Fig. 6.2. C, connection to sub-tegumental cell; M, muscle; P, pit; T, tegument; TOM, tegumental outer membrane. (Photograph by courtesy of Professor Dr H. Melhorn.)

move into the veins of the mesenteries or into those surrounding the urinary bladder. The complete cycle, from egg release to full sexual maturity, may take several weeks.

Once mature the worms survive for a considerable time, possibly as long as five or six years, during which they release a very large number of eggs (*S. mansoni* 300 per day; *S. japonicum* 3000 per day). Eggs are laid by the females into small diameter vessels and become trapped by the elasticity of the vessel wall. Their passage from the vessel, across the tissues to the lumen of the intestine or bladder, is dependent upon enzymes released from the miracidia, which develop precociously within the eggs. As will be seen later, it is this phase of the life cycle that is responsible for the severe pathology associated with schistosomiasis.

6.2   Schistosomiasis

Infections with schistosomes are widely distributed in tropical and sub-tropical regions, their distribution being limited by the climatic conditions appropriate for the survival of the snail hosts. Human schistosomiasis, which affects some 300 million people, is caused primarily by three species, *S. mansoni* (Africa, Central and South America), *S. japonicum* (Asia), and *S. haematobium* (Africa). In the first two species the adult worms live in the mesenteric veins of the intestine; *S. haematobium* inhabits the plexus of veins around the bladder. Several species, notably *S. bovis* and *S. mattei*, cause disease and severe economic loss in cattle and sheep. In all hosts schistosomiasis is a chronic and insidious disease, producing long-term, debilitating pathology.

6.3   Immune responses

As may be expected from the fact that the host is parasitized for long periods by relatively large and reproductively prolific parasites, infection is associated with a wide variety of immune responses. Initial penetration of the skin produces little reaction, but in repeated infections there may be local hypersensitivity responses. (In man such responses are much more pronounced after infection by the cercariae of avian schistosomes. These are unable to develop further, die in the skin and give rise to a violent dermatitis – swimmers' itch.) Early development is sometimes associated with acute allergic reactions, but often the first responses that become evident are those associated with the production of eggs. The constant release of potent immunogens from the eggs produces a strong cell-mediated immunity and this leads to pronounced immunopathological changes (see p. 119). The adult worms are themselves not directly pathogenic, but they are strongly immunogenic. Antigenic material is released from the tegument, from the intestine, and is also released during metabolism. A variety of antibody responses is made to these antigens, including marked reaginic responses.

6.4   Protective immunity

Clear-cut evidence for acquired immunity and resistance to infection has only recently been demonstrated in humans. The fact that individuals do survive in areas with high transmission rates, and observations that infection levels plateau after the second decade of life, have always been taken as circumstantial evidence for such immunity, but it has not been possible to rule out the argument that these patterns of infection reflect behavioural changes resulting in decreased water contact rather than the development of effective immunity. A series of studies with both *S. mansoni* and *S. haematobium*, in which detailed records of water contact have been correlated with levels of infection before and after effective chemotherapy, have shown quite clearly that individuals can maintain low levels of infection despite extensive exposure to waters known to harbour infected snails (Fig. 6.4). It is equally clear from these studies that the ability to express such resistance develops slowly despite repeated infection, and that levels of resistance vary considerably between individuals. The data from these immunoepidemiological studies, together with observations from *in vitro* experiments and from animal models, are at last providing clear insights into the nature of immunity to schistosomes.

In animal models there is clear evidence of immunity, the precise form of immunity depending upon the species of host used. With *S. mansoni*, for example, rhesus monkeys express very substantial immunity on reinfection, but primary infections survive for many months; baboons show less resistance to challenge infections. In the rat, primary infections are themselves terminated by immunity and there is strong resistance to reinfection. Mice however are closer to the rhesus monkey in their overall responses to infection.

Some of the most important experimental studies of the mechanisms underlying immunity to schistosomes were carried out by Smithers and Terry during the 1960s. The results they obtained have exerted a very significant influence upon subsequent approaches to this question. In summary, their work with *S. mansoni* showed that:

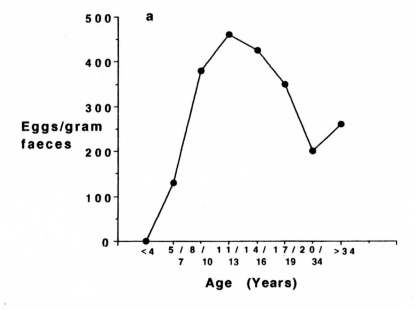

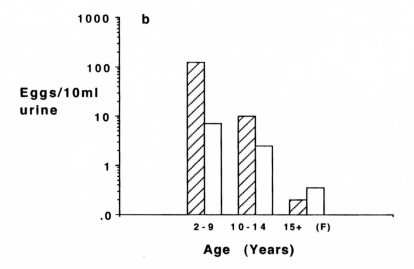

Fig. 6.4 Profiles of infection with schistosomes in populations living in endemic areas. (a) Infection with *S. mansoni* in Brazil, showing age distribution of egg counts, adjusted for degree of water contact and sex of person. (Data from Abel *et al.*, 1991, *American Journal of Human Genetics*, **48**, 959). (b) Relation between degree of exposure to infection and intensity of reinfection with *S. haematobium* in The Gambia. (Data from Wilkins *et al.*, 1987, *Transactions of the Royal Society of Tropical Medicine and Hygiene*, **81**, 29.) ▨ high exposure; □ medium exposure; F, females only

(*a*) Rhesus monkeys were able to destroy the majority of worms developing from a challenge infection, but did not eliminate the adults that had established from the initial infection.

(*b*) Immunity to infection could be stimulated by exposure to large numbers of attenuated (irradiated) cercariae or by transplantation of adult worms directly into the vascular system.

(*c*) Immunity to reinfection appeared to act against the larval stages. Because this immunity could be stimulated by adult worms, the two stages must express common antigens, yet the adults were unaffected by responses that destroyed the larvae.

In 1969 Smithers and Terry proposed the term *concomitant immunity* to describe this paradoxical situation in rhesus monkeys. This term was taken from tumour immunology, where it was used to describe the fact that animals bearing certain primary tumours were able to destroy new tumours of the same kind but were unable to control the inital tumour. It is now recognized that concomitant immunity to schistosomes exists in other hosts and subsequent work has elucidated much about the ways in which adult worms evade the immune response and in which developing larvae are killed. These two aspects will be discussed separately.

6.4.1 Evasion of immunity by adult worms

At the time that the concept of concomitant immunity in schistosomiasis was put forward it was already known that, in homogenates of adult worms, there were antigens that were indistinguishable from those of the host used for the infection. It was realized that the existence of such common antigens would help to reduce antigenic disparity between host and parasite and thereby decrease the impact of immune responses. Smithers and Terry provided direct evidence to support this proposition by carrying out the experiments shown in Fig. 6.5. Immunization against mouse antigens was achieved by injecting monkeys with homogenized liver and spleen cells, with RBC or with serum proteins. When worms were transferred from monkey donors survival was essentially similar whether the recipient had been immunized against mouse antigens or not. 'Mouse' worms survived after transfer into normal monkeys but required about 6 weeks to regain pre-transfer levels of egg output. In monkeys immunized against mouse antigens the majority of 'mouse' worms died within 25 hours and all were dead after 44 hours.

It was also shown in these experiments that the immunity operative against

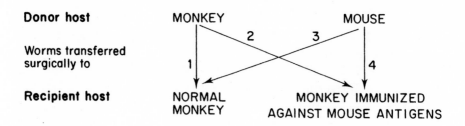

Donor host

Worms transferred
surgically to

Recipient host

Outcome of transfer:

1. MONKEY———▶ NORMAL MONKEY. Worms survived, no interruption
   of egg laying.

2. MONKEY———▶ ANTI-MOUSE MONKEY. Worms survived, no interruption
   of egg laying.

3. MOUSE——▶ NORMAL MONKEY. Worms survived, egg laying interrupted
   for 5-6 weeks.

4. MOUSE——▶ ANTI-MOUSE MONKEY. All worms killed within 44 hours.

Fig. 6.5 Protocol of experiments used to demonstrate presence of host antigens on
the tegument of adult *Schistosoma mansoni*. (Data from Smithers *et al.*, 1969,
Proceedings of the Royal Society B, **171**, 483.)

'mouse' worms in anti-mouse monkeys was transferable with serum, that
immune damage occurred at the surface of the worm, and that the susceptibility of 'mouse' worms to the immune environment of anti-mouse monkeys
was lost if they were first passaged through a normal monkey. After one week
the 'mouse' worms become monkey adapted and were not killed after transfer
into immunized recipients.

It follows from these results that worms acquire from their hosts molecules, called by Smithers and Terry 'host antigens', and have the ability to
turn over these molecules at a relatively rapid rate, i.e. they can be replaced
within one week. The host antigens are present at the surface of the worm,
as can be seen by their interaction with antibodies in the immunized recipient. From the fact that adult worms do stimulate specific and potentially
destructive immune responses, but are not themselves affected, it may be
inferred that host antigens play a protective role, disguising the surface of

the worm so that it is seen as self and thus preventing interaction with host immune mediators.

This view of the way in which adult schistosomes evade the immune response has now been somewhat modified, but basically remains an acceptable explanation. Support for the acquisition of host antigens was provided by *in vitro* experimentation in which it was shown that schistosomula cultured in media together with human RBC would acquire glycolipid antigens of the blood groups A and B on their tegumental surface (glycoprotein antigens were not acquired). If such larvae were subsequently transferred into monkeys that had been immunized against these blood group antigens the larvae were killed rapidly. It is now known that many other molecules of host origin can be acquired, including histocompatibility antigens, immunoglobulins, and skin intercellular substance antigens. The tegument has receptors for the Fc portion of immunoglobulin and for complement components (C1q) and there is good evidence that a host-like α_2-macroglobulin can be actively synthesized. Collectively, these tegumental molecules must play a major role in protecting worms from host immunity. This role is emphasized by the fact that the use of drugs that disrupt the tegument directly (e.g. praziquantel) significantly enhances the protective activity of anti-worm antibodies. However, disguise is not the only protective strategy used. The surface of the tegument is constantly replaced by membrane formed in deeper-lying tissues and this turnover is accelerated when the surface is subjected to antibody and complement-mediated attack. During larval development the schistosomula undergo tegumental changes in lipid constitution which make them intrinsically less susceptible to damage. Finally, schistosomes employ a variety of immunomodulatory strategies that reduce, inhibit or divert host protective responses.

6.4.2   Targets of immunity

Although in general adult worms appear well protected, they can nevertheless be affected by immune responses generated by infection. Field data from human infections suggest that immunity can influence the adult stage in terms of both worm survival and egg output, and this picture is seen more clearly in experimental models. In rhesus monkeys, for example, the egg output of worms maturing from a primary infection with *S. mansoni* reaches a peak after 8 to 12 weeks and then falls rapidly to a much lower level; adult worms may also be lost. A similar situation can occur in baboons infected with *S. mansoni* or *S. haematobium*. In rats, the majority of adult *S. mansoni* are eliminated between the 4th and 8th week of infection. However, despite this evidence for effective anti-adult immunity, it seems clear that the most important targets for effector mechanisms are the early schistosomula that develop from challenge infections. The nature of the effectors and their interaction with larvae have been studied both *in vitro* and *in vivo*.

6.4.3   *In vitro* studies

Although the direct relevance of data derived from *in vitro* studies to events occurring *in vivo* can be questioned, *in vitro* work has helped to lay the foundation for much of our current understanding of effector mechanisms. Schistosomula can be prepared from cercaria by allowing them to penetrate through a skin membrane (skin-transformed schistosomula) or by direct mechanical agitation (mechanically transformed schistosomula). The larvae obtained can then be maintained for long periods in relatively simple media. Viability can be checked visually, by measuring uptake of vital dyes, by release of isotope markers, or by injection into experimental hosts.

A variety of effector mechanisms is known to act against larvae *in vitro*, almost all of which involve cooperative interactions between cells and antibodies (Fig. 6.6). Larvae can also be killed by complement-mediated lysis. The interactions of effector cells with opsonized larvae occur through membrane Fc and/or C3b receptors, but cells can also be armed specifically by antibody or antigen/antibody complexes. Although a number of cells can act as killer cells *in vitro*, the two most important appear to be the eosinophil and macrophage. It is striking that neither natural killer cells nor cytotoxic T cells appear to be able to kill schistosomula, even though both are effective in tumour cell destruction, a partially analogous target. The failure of T cells to kill larvae is even more remarkable given that the parasites can acquire host MHC molecules on the tegument and therefore, potentially at least, have both the self and non-self components necessary to act as targets, although, of course, these acquired MHC molecules are unlikely to present antigens correctly.

6.4.4   Eosinophils

Adherence of eosinophils to larvae can be mediated in several different ways. Where larvae have been been exposed to specific antibody, and complement has been activated, interaction involves both Fc and C3b receptors. Under

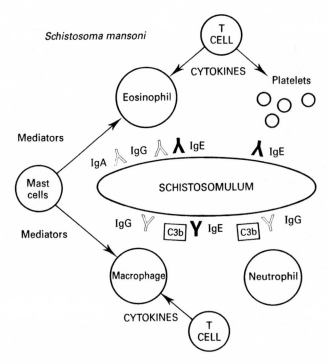

Fig. 6.6 Antibodies and cells involved in antibody-dependent cellular cytotoxicity (ADCC) reactions against the schistosomulum stage of schistosome parasites. IgA also mediates ADCC by eosinophils. (Adapted from Vignali *et al.*, 1989, *Immunology Today*, **10**, 410–16.)

experimental conditions it can be shown that adherence can also occur via the FcR alone, if only antibody is present, or by the C3bR alone, as the tegument activates complement through the alternative pathway. Eosinophils also bind IgE molecules through their low-affinity FcεII receptors, and this allows them to adhere to larvae by antigen-specific interaction through the Fab portions of the IgE.

During infection, at least in the rat, where there have been detailed studies, ADCC involving eosinophils is mediated initially through IgG2a antibodies. These are produced during the first 6 weeks or so of infection, after this time IgE antibodies predominate. The IgG antibodies react with tegumental antigens present on both larval and adult worms, thus immunity to larvae can be promoted by antigens from the adult stage. IgG2a antibodies also attach to mast cells, and antibody-mediated eosinophil killing of schistosomula is enhanced in the presence of these cells, this being associated with the release

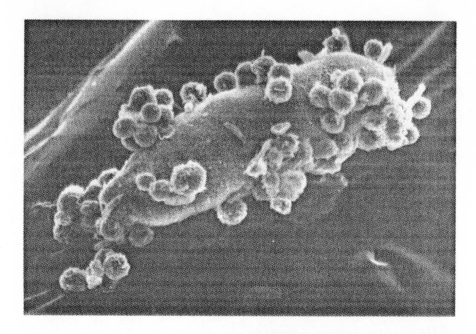

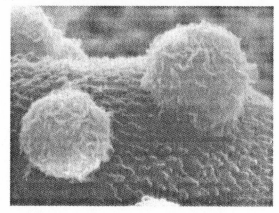

Fig. 6.7   Scanning electron micrographs of eosinophils adhering to the tegument of antibody coated schistosomula larvae of *Schistosoma mansoni*. (Photographs by courtesy of Dr D.J. McLaren.)

of the chemotactic tetrapeptides of ECF-A (eosinophil chemotactic factor of anaphylaxis). Recently, it has been shown that eosinophil-mediated ADCC can also involve IgA antibody.

After adherence to the larvae, eosinophils flatten out and make intimate contact with the tegumental surface (Fig. 6.7). The secretion granules of the cell accumulate adjacent to the point of contact and fuse to form vacuoles. These then fuse with the cell membrane and the contents are released onto the worm (Fig. 6.8a,b). The granules contain a variety of mediators, including enzymes such as peroxidases and phospholipase B, which can damage the tegument. They also contain large, characteristically shaped crystalline bodies whose major component is a strongly basic protein. This major basic protein (MBP) is also released and is potently destructive. Larvae are damaged even at MBP concentrations as low as 2×10^{-5} M. Damage, however caused, occurs initially at the outer bilayer of the tegumental membrane and then at the inner bilayer. These changes alter the permeability of the membranes and the tegument becomes vacuolated. Eosinophils can then actively invade the tegument and strip it away from the underlying muscles (Fig. 6.8c).

6.4.5   Macrophages

Macrophage-mediated killing appears to operate through two distinct mechanisms. One is non-specific and antibody-independent and is mediated by macrophages that have been activated, e.g. by IFN-γ released from antigen-stimulated T cells. This form of killing involves the production of toxic metabolites, including nitric oxide. The other mechanism is specific and involves anti-parasite IgE. This arms the macrophage rather than opsonizes the larva, allowing cells to adhere and release enzymes onto the tegument. IgG antibodies appear to play no role in specific killing by macrophages, indeed cleavage of anti-parasite IgG by parasite enzymes releases the tri-peptide threonine-lysine-proline, which can interfere with IgE-dependent cytotoxicity.

6.4.6   Schistosomula as targets for effector mechanisms

In vitro studies have shown that ADCC killing mechanisms are effective only when the target schistosomula are young; after a few days their susceptibility is much reduced. This rapid development of insusceptibility is essential to the survival of the parasite, even during primary infections. Although antibodies are not present, surface activation of complement via the alternative pathway can result in C3b-mediated cell adherence. Loss of susceptibility is correlated with altered surface antigenicity, so that antibody binding and classical complement activation are reduced, but at the same time the tegument shows an inherently greater resistance to effector-mediated damage. This was elegantly demonstrated some years ago by making older larvae artificially

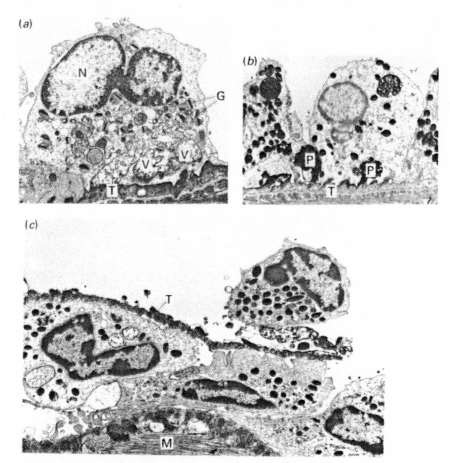

Fig. 6.8   Eosinophil-mediated cytotoxicity against schistosomula larvae of
Schistosoma mansoni. (a) Close adherence of eosinophil, showing fusion of granules
(G) to form secreting vacuoles (V) from which mediators will be released onto the
tegument (T). N, eosinophil nucleus. (b) Release of peroxidase material (P) from
secreting vacuoles onto tegument (T). (c) Stripping of tegument (T) from underlying
muscles (M) by invading cells. ((a) from McLaren *et al.*, 1978, *Parasitology,* **77**, 313; (b)
and (c) from McLaren, 1980, *Schistosoma mansoni: The Parasite Surface in Relation to
Host Immunity*, Research Studies Press, John Wiley. By permission of the author and
publishers.)

immunogenic (by linking haptens to their tegument) and then exposing them
to anti-hapten antibody, complement and eosinophils *in vitro*. (Table 6.1).
Larvae taken from the lungs after 5 days *in vivo* were more than five times more
resistant to killing than skin-transformed schistosomula. The reasons for this
increased resistance to tegumental damage are not fully understood. There is

Table 6.1 *Resistance of lung-stage larvae of* Schistosoma mansoni *to antibody-mediated killing by eosinophils*

Larvae	% of larvae with >5 adherent eosinophils	% degranulated eosinophils on larvae at 18 h	% larvae killed after 36 h
3 hours old skin stage	88.8	53.3	80.5
5 days-old lung stage	100.0	63.3	15.2

Notes:
Larvae were haptenated by coupling trinitrophenyl (TNP) to the tegumental surface, then incubated at 37°C *in vitro* with rabbit anti-TNP antisera and purified human eosinophils.
Source: (Data from Moser *et al.*, 1980, *Journal of Experimental Medicine* **152**, 41.)

evidence that exposure to human sera triggers biochemical changes in the tegument, one result of which is uptake of neutral lipids, but increased resistance is seen even when larvae are cultured in serum-free media.

The evidence that schistosomula can be killed by immune-mediated mechanisms and are a major target of immunity *in vivo* has focused attention on the antigens likely to be relevant to protective responses. Although non-surface antigens are also involved, many are expressed at the tegumental surface (Table 6.2). The biological roles of most of these molecules are unknown, although it can be suggested that they may play roles in signal transduction, nutrient uptake or defence against host immunity. In young skin-stage larvae many of these surface molecules are glycosylated. More than 90% of the surface epitopes of schistosomula are associated with the carbohydrates of the 17, 38 and > 220 kDa antigens, and these are cross-reactive with antigens present in the eggs, providing a molecular basis for the phenomenon of blocking antibodies (see below). Others (e.g. the 18–32 kDa molecules) cross-react with the adult stage and may therefore be involved in the immune responses underlying concomitant immunity.

6.4.7. Other targets

Although schistosomula are a major target for protective responses, there is good evidence that later (post-lung) stages and even adult worms can also be

Table 6.2 *Surface antigens of schistosomes*

Molecular weight (kDa)	Epitope	Schistosomulum 3 hours	Schistosomulum 5 days	Adult worms	Eggs
> 200	CHO	+	−	−	+
38	CHO	+	−	−	+
38	Pep	+	−	−	−
32	Pep	+	+	+	−
25	Pep	−	+	+	−
20	Pep	+	+	+	−
18	Pep	+	+	+	+
17	CHO	+	−	−	+
15	Pep	+	+	+	−
8	Pep	+	+	+	−

Notes:
CHO, carbohydrate; Pep, peptide.
Source: (Taken from Simpson, 1990, *Parasitology Today*, **6**, 40 and Dunne, 1990, *Parasitology Today*, **6**, 45.)

affected by host immunity. A number of the antigens concerned in immune responses are shared by all life cycle stages but, as shown in Table 6.2, some are restricted in their occurrence. In addition to the surface antigens already described, molecules that are located within or below the tegument can also play an important role in immunity and a number have been defined and character-ized. Some of these are of unknown function, e.g. the 22 kDa antigen of *S. mansoni*, responses to which correlate with resistance in human populations, whereas others have clear biological roles, e.g. paramyosin and other muscle pro-teins, glutathione S-transferases and several other enzymes. A striking observa-tion has been the identification of antigens that cross-react with molecules found in molluscs, responses to which can result in enhanced resistance. All stages of the life cycle, for example, express a surface glycoprotein that cross-reacts with keyhole limpet haemocyanin (KLH). Rats immunized against KLH showed 50–70% protection against subsequent challenge with *S. mansoni*.

6.4.8   *In vivo* studies

Immunoepidemiological studies in humans have pointed to a positive correla-tion between anti-worm IgE responses and resistance to reinfection. Using human material, it has been shown *in vitro* that eosinophils, macrophages and platelets are all able to kill schistosomula in the presence of this isotype. There is therefore strong circumstantial evidence that IgE-mediated ADCC may be an important component of protective immunity, although it is certain that other mechanisms must also be involved. Work with small animal models (primarily mice and rats and using *S. mansoni*) has provided a much greater insight into *in vivo* protective responses, but it is clear that there are considerable differences between each model and it is debateable which, if any, is directly analogous to humans. These differences concern the timing of responses, the components involved, and the relationship between immunity generated by infection and immunity generated by vaccination. There are also differences in the sites at which parasites are killed in immune hosts, as has been shown by detailed radiolabelling and tracking studies using autoradiography.

6.4.8.1 Rats – immunity from infection The rat is a relatively non-permissive host, in that immunity is expressed against worms from a primary infection. Adults may fail to reproduce and are eliminated quickly. Experience of a primary infection gives immunity to challenge, and the lung appears to be a major site of parasite destruction. This immunity is T cell-dependent. Early studies involving passive transfer of antisera, and work with rats treated with anti-μ chain antibody from birth, and thus incapable of immunoglobulin production, emphasized the importance of antibody in immunity. Transfer of immune serum into normal rats can provide substantial protection against subsequent infection (a 50% reduction in lung stage larvae) but this effect is dependent upon the availability of inflammatory cell populations and is lost if rats are irradiated prior to transfer. The protective activity of immune sera is associated with the IgG2a and IgE isotypes, as has been shown by selective isotype depletion, and substantial immunity has been transferred with a monoclonal IgG2a antibody directed against tegumental antigens. *In vitro* data suggest that these antibodies act through ADCC mechanisms in which eosinophils play a major role and this is supported by experiments in which immunity has been enhanced by direct transfer of eosinophils, but macrophages also contribute to immunity. Recent studies, in which rats have been vaccinated with a recombinant antigen (Sm28GST) have confirmed the involvement of IgE in immunity, but have also shown that IgA antibody can mediate protective eosinophil-dependent cytotoxicity

6.4.8.2 *Mice – immunity from vaccination with irradiated larvae.* The mouse is a permissive host, allowing reproducing populations of adult worms to establish and survive. Primary infections lead to the development of extensive hepatic pathology following the formation of granulomata around trapped eggs (see below). Challenge infections given to such mice give low levels of adult worm establishment, but it is now recognized that this 'immunity' is in part a consequence of the formation of portal–caval anastomoses (connections between blood vessels) and the diversion of developing worms away from the liver. Mice can, however, be immunized by exposure to irradiation-attenuated cercariae or schistosomula, which do not mature into egg-producing adults, and immunity can then be studied in the absence of pathology.

Vaccine-induced immunity is T cell-dependent and it is clear that the cells involved are the CD4$^+$T helper (Th) subset. Both ADCC (involving IgG but not IgE antibodies) and T cell activation of macrophages have been implicated as effector mechanisms, but complement appears not to play a role. There has been a great deal of debate about where immunity acts. Challenge larvae may be affected shortly after penetration into the skin but the lung appears to be a more important site and has been the focus of recent studies. The mechanisms operative in each location are different. Eosinophils and neutrophils have been implicated in skin responses, whereas macrophages play the major role in the lungs.

Detailed studies of the lung-stage response show that Th1 cells are critical. Following antigen stimulation these cells release IFN-γ, activating macrophages that can then interact with schistosomula, and other cytokines that initiate the development of inflammatory foci in which larvae are trapped. Immunity is much reduced if vaccinated mice are treated with an antibody against IFN-γ (Table 6.3). Not surprisingly, antibodies against IL-4 and IL-5 have little effect. Precisely how macrophages damage or kill larvae is uncertain, but production of reactive oxygen and nitrogen intermediates, and release of TNF, are almost certainly involved.

6.4.9   Interference with immunity

One of the long-standing puzzles in schistosome immunology is the slow development of immunity in human populations exposed to infection. Adult worms, of course, can evade immunity in several ways, but it would seem reasonable to assume that, after a period of infection and immune stimulation, protective responses against the vulnerable schistosomula would prevent or reduce reinfection. Epidemiological studies show that this does happen but it may take 10 to 15 years, depending upon the intensity of transmission. Serological analyses of school-age individuals in large-scale reinfection studies (where infected populations in endemic areas are given effective chemotherapy

Table 6.3 *Effects of anticytokine antibody treatment on resistance to challenge in mice vaccinated against* S. mansoni *by exposure to irradiated larvae*

Treatment	Level of resistance (as % of worms in controls)	
	Vaccinated mice	Vaccinated + antibody
**anti-IFNγ	67%	8%
*anti-IL-4	67%	70%
*anti-IL-5	74%	69%

Notes:
Mice were vaccinated by exposure to 500 irradiated cercariae and challenged one month later. Level of resistance is calculated from the numbers of adult worms recovered from the challenge.
Source: (Data from *Sher *et al.*, 1990. *Journal of Immunology*, **145**, 3911 and **Smythies *et al.*, 1992, *Journal of Immunology*, **149**, 3654.)

and the rate of parasite contact and reinfection monitored) suggest that immune status is to a considerable extent determined by the balance of anti-parasite isotypes produced, particularly IgG4 and IgE (Fig. 6.9). Reinfection was significantly reduced in individuals with high anti-adult IgE and more likely in those with high IgG4. The explanation for these observations lies in the fact that IgG4 antibodies are known to block eosinophil-associated ADCC reactions to schistosomula and may well interfere with other effector mechanisms (e.g. mast cell degranulation). The concept of blocking antibodies in the schistosome–host relationship is well-documented in *in vitro* models. In humans, it has been shown that IgM antibodies to particular epitopes on schistosomal surface antigens effectively block the binding of IgG isotype (ADCC-promoting) antibodies to the same target. In rats, monoclonal IgG2c antibodies block eosinophil killing dependent on monoclonal IgG2a antibodies and in mice, monoclonal IgM antibodies block IgG-mediated ADCC. Blocking antibodies of this kind often show specificity for epitopes (particularly carbohydrate epitopes) that are present both on the larvae and in eggs. These cross-reacting antibodies raised in response to an egg-producing adult infection thus decrease host immunity to invading larval stages.

Synthesis of particular isotypes by B cells is regulated by T cell cytokines. Work in the mouse has shown that the onset of egg production in schistosome infections is associated with a switch from a predominantly Th1 response to a Th2 response, a down-regulation of the cytokines IFN-γ and IL-2 and an increase in production of IL-4, IL-5 and IL-10 (Fig. 6.10). Many

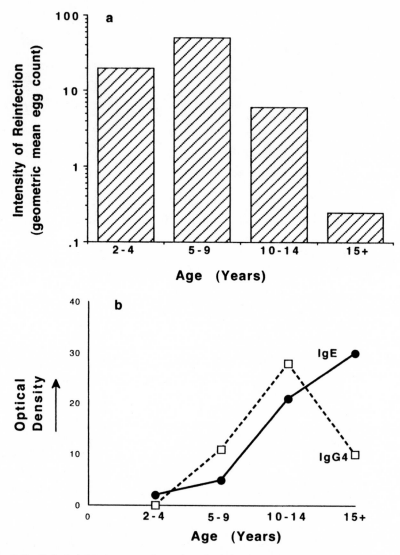

Fig. 6.9 Relationship between intensity of reinfection with *Schistosoma haematobium* 15 months after chemotherapy and IgE and IgG4 antibody responses to worm antigens. (Data from Hagan *et al.*, 1991, *Nature*, **349**, 243.)

of the inflammatory events that occur at this time can be linked to this pattern of cytokine release, and the cytokine balance also influences the pattern of antibody isotype production. An important question, still to be resolved, is whether a similar sequence of events occur in humans; certainly some of the available evidence seems to suggest that it does.

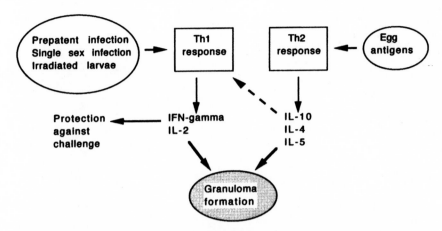

Fig. 6.10 Interrelationships of T helper subsets with protective and pathological responses to schistosome infection. Before egg production occurs, the immune response is polarized towards the Th1 subset with release of IFN-γ and IL-2; after egg production induction of Th2 responses leads to down-regulation of Th1 (via IL-4 and IL-10) and reduced protection. Both sets of cytokines are relevant to granuloma formation.

6.5 Immunopathology

Schistosomiasis is one of several parasitic diseases in which the pathology associated with infection is caused not by the direct activities of the parasite but by the immunological and inflammatory responses of the host. There are a number of well-defined immunopathological symptoms associated with infection, including the dermatitis that may follow cercarial penetration, the acute allergic phase caused by parasite migration through the lungs, the later allergic phase (Katayama fever) seen in *S. japonicum* infections, and immune complex disease. The most intensively studied immunopathological reactions are those associated with chronic *S. mansoni* infections and which are responsible for the gross changes seen in the liver. The close similarity between schistosomiasis in man and that in mice has allowed the underlying causes of this pathology to be studied in great detail.

The intended fate of the eggs released from female worms in the mesenteric vessels is their movement into the tissues and lumen of the intestine and their ultimate release from the host in faecal material. Many eggs, however, do not remain localized in the blood vessels around the intestine, but are swept away by the blood. They enter the hepatic portal vein, pass into the liver and then become trapped in pre-sinusoidal venules. In chronically infected hosts,

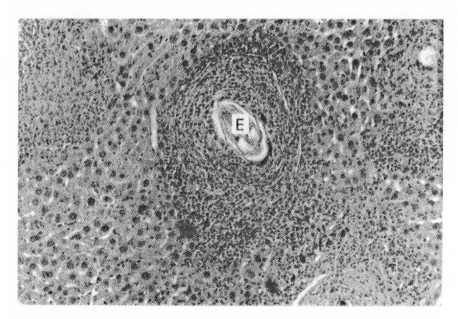

Fig. 6.11   Granuloma around the egg (E) of *Schistosoma mansoni* in the liver of a chronically infected mouse.

long-term exposure to the potent immunogens released by the miracidia within the eggs elicits a state of delayed hypersensitivity and each egg becomes the focus for the production of a granuloma.

The antigens involved in this response are secreted by the miracidia (soluble egg antigens – SEA) and released through pores in the eggshell. SEA is a complex mixture of proteins, glycoproteins, carbohydrates and glyco-lipids. Certain components of these antigens are recognized by T cells, which then play a key role in the initiation, development and subsequent modulation of the granulomata. Antibodies are made to a much wider range of SEA components. The T cells involved are T helper cells, and cytokines released from them, and from other cells, results in the focal accumulation of lymphocytes, macrophages, eosinophils and other inflammatory cells around each egg, producing an inflammatory focus that may reach about 400 μm in diameter (Fig. 6.11). The initial T cell response appears to involve Th1 cells, although this has been disputed. Release of cytokines, particularly TNF, from macrophages activated by this response is then responsible for the initiation of granuloma formation, but SEA subsequently appears to promote the selective development of Th2 cells, whose release of the cytokine IL-5 is responsible for the accumulation of eosinophils. Granuloma cells also release factors that stimulate fibroblast activity, production of collagen by these cells playing a major role in the development of fibrosis in the infected liver.

The larvae in eggs that become surrounded by granulomata are killed, a process in which eosinophils play an important role. Granuloma formation can therefore be considered as a host-protective mechanism, but one that carries a considerable cost in terms of pathology, given that the host has limited ability to prevent egg release by female worms. However, the immune response that leads to granuloma formation also protects the host from the effects of hepatotoxins released by the miracidia within the eggs. In T cell-deficient mice, which cannot form granulomata or antibodies to SEA, there is necrosis of liver tissue around the trapped eggs. A fascinating aspect of granuloma formation is that, in the intestine the immune response to the eggs appears to promote their movement through the tissues allowing more of them to reach the intestinal lumen and thus pass into the outside world. The parasite is therefore exploiting for its own ends a response mounted by the host for its protection. A similar exploitation has been uncovered by the observation that host production of the cytokine TNF may stimulate egg release by the female schistosome. Since TNF is toxic for schistosomula, its production may also serve the interests of the adult parasites by preventing superinfection.

6.5.1   Immunomodulation of granulomata

As hosts become sensitized to SEA the inflammatory response to trapped eggs becomes more intense, and granuloma formation is enhanced. With time, however, the size of existing granulomata and of new granulomata decreases. This modulation is the result of complex immunoregulatory influences, primarily dependent upon T cell and cytokine activity. Transfer of cells from mice infected for 20 to 30 weeks, which show modulation, into mice infected for 6 weeks, which do not, results in a significant reduction in the size of newly formed granulomata. One factor contributing to modulation may be a reduction in the efficiency of granuloma macrophages as antigen-presenting cells.

7

Gastrointestinal nematodes
Immunity within the intestine

7.1   Introduction

The vertebrate intestine can be considered one of the major ancestral sites for parasites. In evolution, access to the bodies of vertebrate hosts would have been achieved most easily through accidental ingestion. Survival in the intestine would have been favoured by the high plane of nutrition available, and the continuation of the species would have been ensured by the ready exit to the outside world. Intestinal species are, overall, still the commonest, although not the most pathogenic, of all parasites and this is seen very clearly in the nematodes. Although many species parasitize deeper tissues of the body (see Chapter 8) the majority are intestinal, and the intestine has been retained as the site of adult development even when infection occurs through skin penetration and complex tissue migrations have been incorporated into the life cycle (Fig. 7.1).

Intestinal nematodes are ubiquitous parasites of man and domestic animals (Table 7.1). The diseases they cause are rarely fatal, but instead are long-term and debilitating, their effects upon the host originating in pathological alterations of intestinal structure and function. In man such infections are common in countries where climatic and sociological conditions favour transmission, but certain species, notably the pinworm *Enterobius vermicularis*, are equally common in the developed nations of the Western world. In domestic animals gastro-intestinal infections are invariable accompaniments of high density stocking and intensive production, and are responsible for enormous economic losses.

Many studies have shown that the intestine cannot be considered as a single habitat for parasites, but must be viewed as a series of habitats, each with its own distinctive characteristics. These characteristics change longitudinally along the length of the intestine, and also radially, from the lumen to the mucosa. Particular species have preferred locations in the intestine and are capable of active migration to such locations after infection or experimental implantation. Large worms such as *Ascaris* must necessarily live within the lumen, but small species such as the hookworms and trichostronglyes have an intimate association with mucosa and thus experience very different environmental conditions. Some species live within the mucosa itself during their developmental stages, emerging into the lumen when mature; a few remain wholly or partially in the mucosal tissues through their life in the intestine. Of these, *Trichinella spiralis* and species of *Trichuris* are thought to have intracellular locations, penetrating within the cells of the epithelial layer.

7.2 Immune responses in the intestine

At one time it was thought that worms living in the gut lumen were effectively outside the body and could neither initiate nor be affected by immune responses unless they damaged the mucosa and breached what was considered to be an effective barrier. It is now known of course that this view is quite incorrect and that intestinal worms are as much subject to protective immune responses as those living elsewhere in the body. However, there are some important differences in the nature of those responses that operate in the intestine, and it is necessary to outline briefly some of their characteristics before considering particular infections.

7.2.1 Enterocytes and antigen uptake

The gut is the major route of entry for antigens into the body, not only those produced by infectious organisms, but also antigens associated with food, environmental contaminants and the normal bacterial flora. The enterocytes that form the epithelial layer of the mucosa provide a physical barrier, but uptake of antigens across this layer occurs readily, and is increased when the permeability of the junctions between the enterocytes is increased during inflammation. Cytokines released during inflammation also result in expression of Class II MHC on enterocytes, enabling them to function as antigen presenting cells. Antigen uptake also occurs via specialized cells overlying the Peyer's patches. Therefore, despite the fact that many potential antigens are

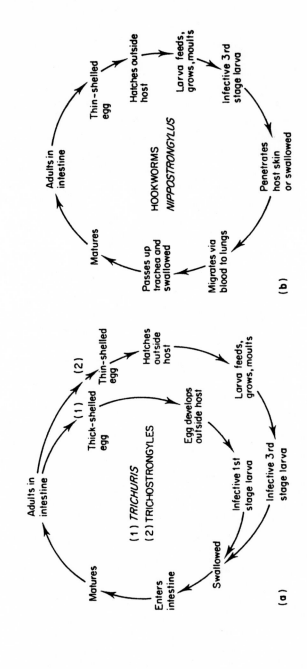

Fig. 7.1 Representative life cycles of intestinal nematodes.

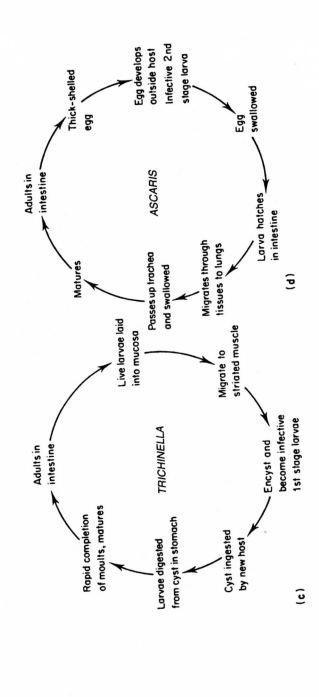

Adults in intestine

Thick-shelled egg

Egg develops outside host
Infective 2nd stage larva

ASCARIS

Matures

Passes up trachea and swallowed

Migrates through tissues to lungs

Larva hatches in intestine

Egg swallowed

(d)

Live larvae laid into mucosa

Migrate to striated muscle

Adults in intestine

TRICHINELLA

Encyst and become infective 1st stage larvae

Rapid completion of moults, matures

Larvae digested from cyst in stomach

Cyst ingested by new host

(c)

Fig. 7.1 (cont.)

Table 7.1 *The major gastrointestinal nematode parasites of humans and domestic animals*

Humans	Cattle/Sheep/Pigs
Ascaris	*Ascaris*
Enterobius	*Cooperia*
Hookworms	*Haemonchus*
Ancylostoma	*Nematodirus*
Necator	*Oesophagostomum*
Strongyloides	*Ostertagia*
Trichinella	*Strongyloides*
Trichuris	*Trichostrongylus*

broken down enzymatically, there can be a substantial transport into the mucosa and it is obviously necessary for there to be effective regulatory mechanisms to prevent overloading of the immune system. Antigens can be complexed and trapped by antibodies present in the surface mucus. If they pass this defence and enter the bloodstream they can be filtered out (particularly if complexed) by phagocytic cells in the liver. Intestinal presentation of antigen can also lead to non-responsiveness, and this provides an effective way of preventing excessive immunological activity.

7.2.2   Intestinal immunoglobulins

The major isotype found at the mucosal surface is dimeric IgA, which can be secreted across the enterocytes, or (particularly in rodents) enter with bile after transport across bile duct epithelia. IgA molecules remain intact and functional in the intestinal lumen by virtue of the secretory piece, which protects inter-molecular bonds from enzyme action. IgM is similarly transported across the mucosal epithelium and remains functional in the lumen. IgG isotypes are produced locally, from plasma cells in the lamina propria, and can also enter the intestinal tissues from the blood. Unless the mucosa is inflamed, when both vascular and epithelial permeability increase, relatively little IgG passes into the lumen. Inflammation-induced movement of IgG into the gut has been termed pathotopic potentiation and may be seen as a device to enhance resistance to infection. Unlike IgA and IgM, IgG molecules are rapidly broken down by proteolysis, but Fab fragments may remain intact and functional for some time. IgE is found within the mucosa, some secreted by lamina propria plasma cells, some entering from the circulation. Much of this IgE may be bound to mast cells.

Levels of complement within the mucosa are similar to those in other tissues. Some does enter the intestinal lumen, but it is not clear what function it has there.

7.2.3   Lymphocytes

The lamina propria contains a very large number of both T and B lymphocytes, the latter contributing to the immunoglobulins present in the mucosa and in the lumen. Both the CD4$^+$ and CD8$^+$ subsets are represented in the lamina propria T cells, but the former are the most important in terms of anti-parasite responses. A specialized population of T cells, the intra-epithelial lymphocytes (IEL), is also located within the epithelial layer. Their function in infections is uncertain, although they do have cytotoxic capacity and release cytokines. IEL are CD8$^+$ and a majority carry T cell receptors composed of γ and δ chains rather than the conventional α and β chains. The Peyer's patches, which occur along the length of the intestine, are discrete units of organized lymphoid tissue, separated from the lumen by a single layer of specialized epithelial cells, the M (microfold) cells. M cells can take up and present antigens from the lumen; Peyer's patches can therefore play an important role in initiating responses to antigens present in the intestine

As with systemic lymphocytes, there is a well-defined circulation of cells to and from the intestinal mucosa. Dividing cells from Peyer's patches and the draining mesenteric nodes pass into the thoracic duct lymph and selectively home, in an antigen-independent manner, back to the intestinal and other mucosal surfaces. Many of the B cell blasts in thoracic duct lymph become IgA-secreting plasma cells in the lamina propria; T cell blasts also home to the lamina propria.

7.2.4   Myeloid cells

There are many non-lymphoid effector cells in the normal mucosa and their number increases during parasitic infection. They include natural killer cells, macrophages, neutrophils, eosinophils and basophils as well as a subset of mast cells – the mucosal mast cells. These, though similar to connective tissue mast cells in containing amines and other mediators, and in having high-affinity FcεI receptors for IgE, show several distinct properties and probably have

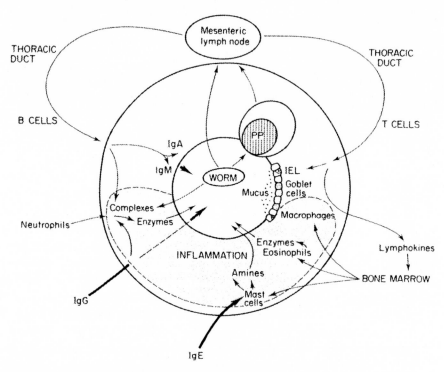

Fig. 7.2   Immune and inflammatory components of protective responses against intestinal worms. IEL, intraepithelial lymphocytes; PP, Peyer's patch. (Adapted from Befus & Bienenstock, 1982, *Progress in Allergy*, **31**, 76.)

a different origin. Infiltration of the mucosa by mast cells, basophils and eosinophils during infection is T cell-dependent and is controlled by cytokine release. Mediators, including cytokines, from all of these myeloid cells, play a major role in the generation of intestinal inflammation. They influence intestinal structure and function, behaviour of lamina propria and epithelial cell populations, and alter the amount and properties of intestinal mucus.

The various ways in which immune and inflammatory reactions can act against intestinal worms is shown in Fig. 7.2.

7.3   Protective immunity against intestinal nematodes

Nematodes present some particular problems for the host's immune reponse, in that they are active organisms and their bodies are covered by a relatively tough external cuticle. Although this is no longer considered to be metabo-

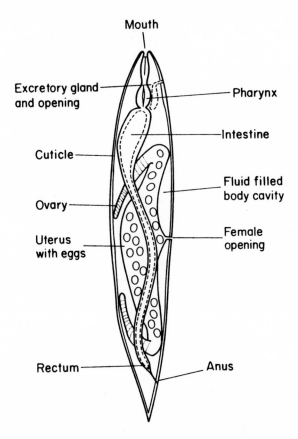

Fig. 7.3   Schematic diagram of a nematode to show major structures and organ systems.

lically and immunologically inert, it nevertheless forms a substantial protective layer, being penetrated only by the mouth, anus or cloaca, female opening, excretory pore and small sensory structures (Fig. 7.3). The cuticle is both antigenic and immunogenic, and can be damaged by immunologically mediated effector mechanisms, but it is doubtful whether responses directed against it play a major role in immunity against intestinal species. It is more likely that protective responses are initiated and expressed against antigens that are released from the body.

   Little is known about protective responses to intestinal nematodes in humans. Epidemiological data suggest that immunity can be acquired as a result of infection, but the fact remains that, typically, worms persist for long periods and reinfection occurs frequently. More is understood of immunity against intestinal worms in sheep and cattle, but most of our detailed knowledge has come from studies made with model systems in rodents. Some use species of nematodes that can infect man or domestic stock (*Trichinella spiralis, Trichostrongylus colubriformis*) or related species (*Strongyloides ratti, Trichuris muris*), others use rodent-specific nematodes that approximate to clinically or economically important species (*Heligmosomoides polygyrus, Nippostrongylus brasiliensis*). In most, infection stimulates responses that lead to expulsion of worms from the intestine during the course of a primary infection ('spontaneous cure'). In all cases, spontaneous cure is immunologically mediated and T cell dependent, as shown by:

- absence of expulsion in T-deficient or T-deprived hosts;
- delayed expulsion in immunosuppressed hosts;
- accelerated expulsion on reinfection;
- transfer of the capacity to express accelerated expulsion with immune serum and/or T cells from immune hosts.

It has been shown that the antigen-specific immune responses stimulated by infection are but one component of the mechanisms in worm expulsion and it is clear that inflammatory events also play a significant role. The importance of inflammation was emphasized as long ago as 1939 by Taliaferro and Sarles in their classic paper describing infections with *Nippostrongylus* in rats. Their view was reinforced by observations made on another form of worm expulsion, 'self-cure' in sheep infected with *Haemonchus contortus*. The original experiments in this system made by Stoll showed that sheep carrying an established infection of *Haemonchus* expelled this infection after a superimposed challenge infection with larval stages (Fig. 7.4) and thereafter were refractory to further challenges for some time. Subsequent studies showed that self-cure was associated with pronounced inflammatory changes in the intestine and raised levels of histamine in the blood, suggesting that immediate hypersensitivity reactions might be involved.

7.3.1   Immunity to *Nippostrongylus*

The involvement of immediate hypersensitivity (IH) in worm expulsion has been most extensively investigated in the rat–*Nippostrongylus* model. The components necessary for intestinal IH (high levels of IgE and presence of mucosal mast cells) are elicited strongly by infection, indeed *Nippostrongylus* not only stimulates high titres of specific anti-worm IgE, it also potentiates

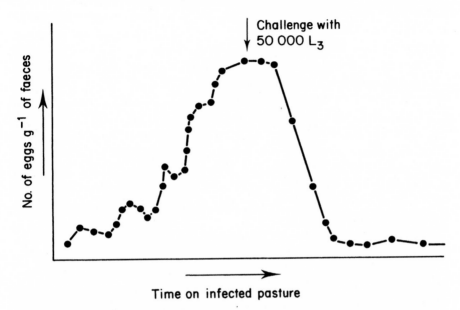

Fig. 7.4   Self cure in sheep infected with *Haemonchus contortus*. Sheep kept on pasture contaminated with infective L_3 were allowed to build up a population of adult worms before being given a heavy L_3 challenge. The egg count dropped dramatically and worms were lost from the intestine.

simultaneously the production of IgE against many other antigens through effects upon the T cell and cytokine regulation of this isotype. Three major hypotheses have been proposed to explain the involvement of IH in worm expulsion:

- Amines released from mast cells damage worms directly.
- Intestinal inflammation resulting from IH produces an environment unsuitable for worm survival.
- Vascular and epithelial permeability, induced by amine release, allow anti-worm (IgG) antibodies to enter the gut lumen (the 'leak-lesion' hypothesis).

There is evidence for and against each of these hypotheses, but no consensus view. Certainly worms are damaged during the course of a primary infection, i.e. egg laying declines, cytopathological changes and lipid accumulations occur in worm tissues and there are alterations in metabolic activity and acetyl-cholinesterase isozyme patterns. During the early phase this damage is reversible and worms survive and resume egg laying after transplantation into new hosts. Eventually damage becomes largely irreversible and worms fail to recover even if transplanted. The cause of this damage is controversial. *In vivo* adminstration of amines can induce similar changes, as can the activity of immune serum in irradiated rats, but 'damage' is also seen when worms are kept *in vitro* under sub-optimal conditions.

Intestinal inflammation can be induced in rats sensitized to a non-parasite antigen by intravenous challenge. When this is done in infected rats there is no enhanced expulsion of worms, implying that inflammation by itself has no effect upon worm survival. Worm loss does occur, however, if rats treated in this way are also injected with immune serum. This observation led to the development of the leak lesion hypothesis, an idea that was strengthened by later work showing that there is a marked increase in mucosal permeability at the time of worm expulsion.

A great deal of controversy has centred around the involvement of IgE and mucosal mast cells in worm expulsion. Necessarily much of the evidence in favour of such an involvement is circumstantial and depends upon evidence that mastocytosis and intestinal permeability correlate in time with worm loss. More recently, evidence has been obtained, by measuring intestinal and serum levels of specific serine proteinases known to be released when mast cells degranulate, that not only do mast cell numbers increase before worm expulsion occurs, the cells also actively release mediators at this time. The correlation between increased enzyme levels and the onset of worm loss is striking (Fig. 7.5). However, none of these correlations is absolute. In lactating rats, for example, mucosal mastocytosis and increased permeability can occur without worm loss, and in certain strains of rats worm loss precedes the rise in mast cells. Significant increases in serum IgE usually follow worm expulsion, but this is not necessarily evidence against a role for specific anti-worm IgE being available at the intestinal level before expulsion takes place, indeed specific IgE is known to be present in the mesenteric lymph node well before it is detectable in serum.

An alternative explanation of worm expulsion is that it does not involve mast cells, but depends on changes in mucosal goblet cells. According to this view, T cell-mediated responses first damage worms and increase goblet cell numbers, then damaged worms induce biochemical changes (altered glycosylation) in mucus, which cause the damaged worms to leave the intestine.

Although *Nippostrongylus* is a natural parasite of the rat, it is also infective to mice. Worm expulsion is more rapid from the mouse and is again associated with mast cell and IgE responses. Normal mice made incapable of IgE synthesis by anti-immunoglobulin treatment or by injection of anti-IL4 antibody can expel *Nippostrongylus* in a normal time scale, suggesting the IgE has no role

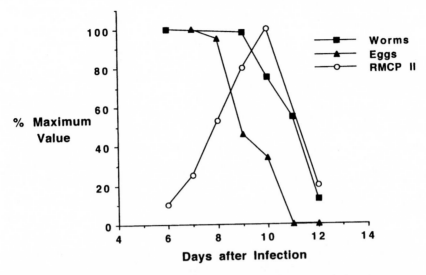

Fig. 7.5 Infection with *Nippostrongylus brasiliensis* in rats. Correlations between numbers of adult worms, faecal egg output and serum levels of rat mast cell protease II (RMCP II) – a mucosal mast cell-specific enzyme released on degranulation. Values for each parameter have been expressed as a percentage of the maximum value recorded. (Data from Woodbury *et al.*, 1984, *Nature*, **312**, 450.)

in worm expulsion. Mice that are genetically deficient in mucosal mast cells (W/W^v strain) do expel worms, but expulsion is delayed. Immune-deficient SCID mice and CD4+ T cell-depleted BALB/c mice do not expel worms, but can do so if given IL-4 to boost their mucosal mast cells; this has no effect in W/W^v mice. These data certainly show the mast cells are not *necessary* for worm expulsion, but do imply that they *can* bring it about. In the absence of mast cells, expulsion may be achieved by changes in goblet cells; in the absence of T cells, it may result from mast cell activity.

7.3.2   Immunity to *Trichostrongylus*

T. colubriformis is a parasite of sheep that has been adapted to the guinea pig. It has a typical trichostrongyle life cycle, and all parasitic stages occur within the gut lumen, close to the mucosa of the small intestine. Primary infections last for two to three weeks before spontaneous cure; challenge infections are expelled more rapidly, at about the time when the L4 stage is present. During infection the intestine is infiltrated by a variety of cells, of which eosinophils, basophils and mast cells are important components. Evidence suggests that amine release from basophils and mast cells plays a major role in immunity. Amine levels in the gut wall and gut lumen rise with the onset of worm loss, amines can also damage worms *in vitro* and induce worm expulsion when given *in vivo*. Conversely, treatment of hosts with amine antagonists or blocking agents significantly delays worm loss. The accumulation of basophils within the mucosa is T cell dependent and is analogous to the accumulation of these cells at sites of cutaneous basophil hypersensitivity (Jones–Mote reactions). Askenase has proposed that there is a complex interplay between T cells, IgE, basophils and other cells in generating these responses. T cells attract basophils into sites of antigen stimulation and induce amine release. The increased vascular permeability that results allows more basophils and T cells to enter the inflamed tissue, and other inflammatory cells, notably eosinophils, are also attracted, leading to the development of a localized inflammatory response.

There are many close similarities between responses to infection with *T. col-ubriformis* in guinea pigs and responses in sheep, in particular concerning the relationship of intestinal inflammatory changes to worm loss. Sheep show marked genetically determined variation in their ability to resist infection, and it is significant that high-responders, i.e. sheep able to control infection well, have more effective inflammatory responses than low-responders, and a greater ability to generate eosinophilia.

7.3.3   Immunity to *Trichinella*

Spontaneous cure of *Trichinella* is a very efficient process in both rats and mice, taking between 10 to 15 days for the worms to be completely removed from the intestine. As with *Nippostrongylus*, worm loss is associated in time with profound inflammatory changes, the most obvious of which are infiltration of the mucosa by mast cells, villous atrophy and crypt hyperplasia, net secretion and accumulation of fluid in the gut lumen, and increased peristalsis.

These are accompanied by a number of more subtle structural, functional and biochemical changes in the epithelial and lamina propria cells of the mucosa. It seems reasonable to suggest that the result of all of these changes is to make the intestine inhospitable for the worm, changing its environment to such a degree that it is no longer able to maintain its preferred position in the small intestine. Several lines of evidence support this suggestion:

- Loss of worms and the onset of inflammation are closely correlated, suppression of inflammation prevents spontaneous cure.

- Worms implanted directly into the gut of a mouse whilst it is responding to a primary, oral infection with *Trichinella* will establish briefly, but are then expelled at the same time as the worms that have induced the inflammation.
- Worms show structural and functional changes as inflammation develops, but are nevertheless expelled alive. Worms recovered during the process of expulsion can survive and reproduce if transplanated into new hosts.
- The inflammatory changes evoked by *Trichinella* will remove from the intestine any other parasites established concurrently.

The inflammatory changes that result in expulsion of *Trichinella* are immune-induced and dependent upon the local activity of a population of CD4$^+$ T cells (primarily Th2 cells) that develops in the lamina propria and draining mesenteric lymph node. These cells cannot bring about worm expulsion themselves, but interact with bone marrow-derived, myeloid cell populations to do so. This interaction is dependent upon release of cytokines that operate both locally and centrally (at the level of the bone marrow) to generate the differentiated cell populations (mast cells, eosinophils) that infiltrate the intestinal wall. A great deal of correlative data supports the idea that mast cells are functionally involved in worm loss, the major lines of evidence being:

- Thymus deficient (nude) mice and mutant mice that cannot mount a mucosal mastocytosis allow *Trichinella* to persist for longer in the intestine than normal mice. Restoration of mast cell responses helps to restore the ability to expel worms.
- As in *N. brasiliensis*, worm loss correlates with an increase of mast cells in the mucosa and with the release of mucosal mast cell-specific proteinase.
- Blocking mast cell development with monoclonal antibody against a stem cell membrane marker (*c-kit*) prevents the normal expulsion of worms (Table 7.2a).

Development of mastocytosis depends upon release of a number of cytokines, primarily IL-3 and IL-9. IL-9 is released from Th2 cells, and Th2 responses characterize *Trichinella* infections in mice. Th2 cells also release the cytokine IL-5, which is necessary for development of eosinophils. Infiltration of the mucosa by eosinophils also accompanies infection, and there has been much speculation that eosinophil-derived enzymes and mediators may also be involved in worm expulsion. However, inhibition of eosinophilia by treating infected mice with an anti-IL-5 monoclonal antibody does not stop worm

Table 7.2 *Effects of depletions of mast cells or eosinophils on immunity to* T. spiralis *in mice*

(a) *Numbers of mucosal mast cells, levels of serum mast cell protease and numbers of adult* T. spiralis *in mice treated with anti-c-kit antibody (ACK-21) before infection with 360 larvae*

Group of mice	No. of mast cells per villus-crypt	Levels of mast cell protease in serum	No. of worms in gut day 10
Infected + control antibody	21.5	17 375 ng/ml	58.8
Infected + ACK-21	0.7	37 ng/ml	186.2

(b) *Numbers of circulating eosinophils and numbers of* T. spiralis *muscle larvae in mice treated with anti-IL5 antibody (TRFK-5) before infection with 350 larvae*

Group of mice	No. of eosinophils per ml blood day 21	No. of muscle larvae per mouse
Infected + control antibody	818	20 433
Infected + TRFK-5	30	22 667

Source: (Data taken from (a) Grencis *et al.*, 1991, *Parasite Immunology*, **15**, 55, and (b) Herndon & Kayes, 1992, *Journal of Immunology*, **149**, 3642.)

expulsion, suggesting that if eosinophils do have a role it is not essential (Table 7.2b).

This focus on inflammatory processes begs the question of whether anti-worm antibodies have any role to play in worm expulsion. In the mouse, passive transfer experiments suggest that antibodies (IgA and IgG isotypes) may interfere with worm growth and reproduction, possibly through an effect upon nutrition, but do not directly cause worm loss during primary infections. It is obvious, however, that if worms are affected early enough they may be less able to maintain themselves in an environment that becomes progressively less suitable for their survival.

Experience of a primary infection with *Trichinella* generates strong immunity to reinfection. In the mouse this is normally expressed as an accelerated version of the primary response, but under certain conditions there can be a very rapid expulsion of challenge larvae within 24 hours. In rats, rapid expulsion is the normal response to reinfection and a variety of mechanisms have been proposed to explain this dramatic loss of worms. One explanation is trapping of larvae in mucus, preventing access to the mucosa and allowing removal of larvae by peristalsis. Antibody in the mucus of immune animals may play a role in this process, perhaps through interaction with antigens present on the cuticle. Passive transfer experiments have confirmed that antibodies can mediate rapid expulsion, but the class of antibody involved (IgG isotypes) and the timing of transfer and worm loss suggest that in this case the antibodies are acting within the mucosa. Certainly some challenge larvae do penetrate into the mucosa, but these are then lost rapidly. The process of rapid expulsion is associated with a number of electrophysiological changes in the epithelial cells of the mucosa, reflected in altered transmembrane potential, flow of ions and movement of water. It has been very clearly shown that these changes are induced by the interaction of parasite antigen with IgE antibodies bound to mucosal mast cells. A further consequence of this interaction is altered secretion of mucus by goblet cells. Rapid expulsion, like spontaneous cure itself, may therefore be the outcome of a complex series of responses to the invading parasite. The rapid loss of other nematodes from immune hosts (e.g. *Nippostrongylus* from rats and *Haemonchus contortus* from sheep) may also involve mucus trapping and physiological changes in the mucosa.

7.4   Antigens involved in immunity

In general, little is known of the antigens that stimulate protective responses against intestinal nematodes. Of the many species studied, more is known of the antigens of *Trichinella spiralis* than any other species.

As in all members of the Trichinelloidea, the oesophagus of *Trichinella* is extremely narrow, embedded in a chain of large cylindrical cells, the stichocytes; collectively the structure forms the stichosome (Fig. 7.6). The cytoplasm of the stichocytes is rich in a variety of membrane-bound granules and there is good evidence that these pass from the stichocytes through small ducts into the lumen of the oesophagus and are released from the mouth. It is presumed that, *in vivo*, this release is associated with penetration into the epithelium and with feeding. This view is supported by the fact that infective muscle larvae release almost all of their granules within the first 24 hours after infection, the granules then being resynthesized.

Stichosomal granules and material secreted during *in vitro* maintenance are highly immunogenic and can elicit responses that reduce worm growth and fecundity and lead to accelerated worm expulsion. Analysis of these antigens shows them to be a complex mixture of proteins and glycoproteins, but a number of molecules (e.g. those of 43 and 50/55 kDa) are immunodominant, largely through their carbohydrate epitopes, and induce strong protective responses. It appears that some of the stichosomal antigens bind to the

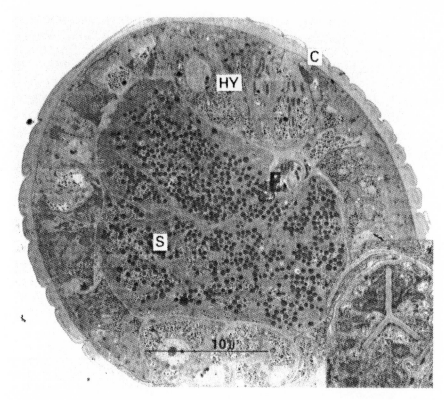

Fig. 7.6 *Trichinella spiralis*. Transmission electron micrograph showing a transverse section through the anterior end of an infective muscle larva. C, cuticle of worm; E, oesophagus; HY, hypodermis; S, stichocyte. Inset shows detail of the oesophagus. (Photograph from Takahashi *et al.*, 1988, *Parasitology Research*, **75**, 42, by permission of the author and publishers.)

cuticle, but the cuticle also expresses its own unique and stage-specific antigens. Those of the newborn larvae may well act as targets for ADCC reactions, but it seems unlikely that cuticular antigens play a significant role in immunity to the intestinal stages.

The existing data we have about *Trichinella* antigens makes it possible to put forward a scheme to explain the sequence of events in the initiation and expression of immunity during a primary infection (Fig. 7.7). Infective larvae are released from the stomach into the upper small intestine, are activated and rapidly penetrate the epithelial layer of the mucosa. Stichosomal antigens are released and generate T cell responses in the lamina propria and draining mesenteric node. T cell and B cell blast cells then migrate back to the small intestine via the thoracic duct. Release of T cell cytokines initiates inflamma-

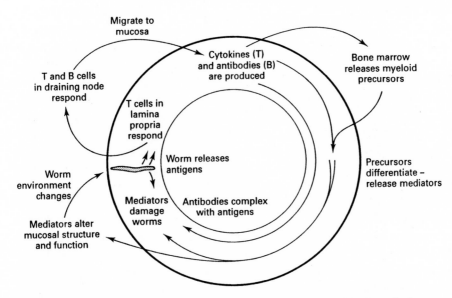

Fig. 7.7   Sequence of immune and inflammatory events initiated by intestinal infection with *Trichinella spiralis*.

tory changes that alter the mucosal environment and induce the differentiation and infiltration of a variety of cells, notably mast cells. The combination of the environmental changes experienced by the worms, damage from inflammatory mediators and the interaction of antibodies with stichosomal products, reduces the viability of the worms, forces them from their preferred niche, and eventually results in their removal by peristalsis.

Newborn larvae of *Trichinella* lack stichosomal antigens, which are synthesized as larvae mature in the muscles. Antigens are released into the muscle cell cytoplasm and appear to initiate the sequence of changes that converts the host cell into the characteristic nurse cell, possibly through a direct interaction with host cell nuclear DNA. This release also maintains production of antibodies, although these appear to have no protective activity against the encysted larvae.

7.5   Chronic infections

Nippostrongylus and *Trichinella* infections are characterized by strong protective immune responses against the worms of a primary infection and by high levels of resistance to reinfection. Clearly, if this was typical of all intestinal nematode infections they would not present the problems that they do. In humans, intestinal nematodes characteristically cause chronic infections and there is little concrete evidence of an effective immunity. Some epidemiological data can be interpreted as showing the existence of protective immunity (e.g. aggregation of infection into small numbers of susceptibles, reduction of intensity of infection with age) but there is certainly no evidence for strong spontaneous cure responses. Many infections in domestic animals are also chronic, even though there is better evidence for the existence of immunity. Thus, although some rodent models provide valuable insights into the ways in which mucosal immunity *can* protect against intestinal nematodes, other experimental systems are necessary to discover why such responses are in many cases inoperative or ineffective.

Failure to generate effective responses may have many explanations and, in the field, it cannot automatically be assumed that chronic infections reflect an absolute inability to develop resistance. Age, health, diet and physiological state all influence the development and expression of immunity as does the level and frequency of infection. Laboratory infections are usually given as a single, large pulse, which would be the exception in the field, where hosts are typically exposed to continuous or interrupted low-level 'trickle' infections. It was shown many years ago that, when *Nippostrongylus* is given in this way, rats may develop large and persistent infections, i.e. the normal spontaneous cure response does not occur (Fig. 7.8). Rats given five larvae per day for 12 weeks accumulated approximately 100 adult worms, 30% of the total given. This is lower than the 50% expected after a single pulse infection, and the worms were stunted, suggesting some resistance. Even when the trickle dose was raised to 50 larvae per day a substantial worm population still accumulated in the intestine. Worms can also be established in immune rats by means of a trickle infection; the worms are stunted and less fecund than normal, but nevertheless survive. Evidence suggests that this mode of infection allows worms to adapt phenotypically to the immune environment, becoming both less immunogenic and less susceptible.

Work with *Nippostrongylus* has also shown clearly that the normal pattern of worm expulsion fails to develop when infections are given to very young rats, i.e. those less than 6 weeks old, and worms may persist into adult life. This situation has significant parallels with worm infections in sheep and may be relevant to some infections in humans, where children are most at risk.

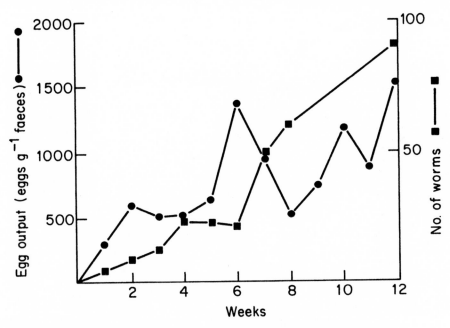

Fig. 7.8   Build-up of infection with *Nippostrongylus brasiliensis* by daily administration of five infective larvae to rats. (Redrawn from Jenkins & Phillipson, 1970, *Parasitology*, **62**, 457.)

7.5.1   Chronic infections in experimental models

In addition to the factors discussed above, genetic and nutritional influences can also affect the host's capacity to respond protectively to infection. Immunity may also be depressed when the host has other infections, including parasites, or is pregnant or lactating. However, even when all the considerations can be ruled out, there still remain species which, at least in certain hosts, seem not to evoke protective responses. Two have provided particularly fruitful experimental models, *Heligmosomoides polygyrus* and *Trichuris muris*; both can be studied in mice.

H. polygyrus has a typical trichostrongyle life cycle, but the larvae spend approximately 8 days within the gut wall before emerging and maturing as adults in the gut lumen. When mice of certain strains (e.g. C57 BL/10) are given a primary infection, the adult worms may persist for up to 10 months, and their eventual loss seems to owe more to senility than to immune expulsion. Challenge infections then develop as successfully as the preceding

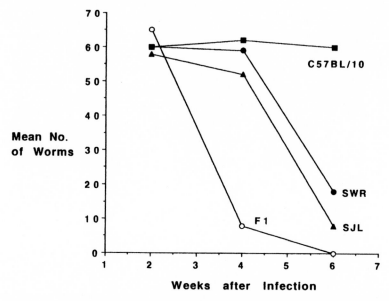

Fig. 7.9 Time course of infection with *Heligmosomoides polygyrus* in three inbred strains of mice – C57BL/10, SWR, SJL and (SWR × SJL)F$_1$ – assessed by numbers of adult worms recovered. (Redrawn from Wahid & Behnke, 1993, **107**, 343.)

primary and do not seem to be greatly affected by immunity. Despite this evidence for long-term survival it is possible to show that infection does elicit significant immune responses, and under appropriate circumstances these responses can be protective. In other mouse strains adult worms are expelled much more quickly, in SWR and SJL mice, for example, within 6 weeks (Fig. 7.9). Strain variation is genetically controlled, and both background and MHC-linked genes are involved. Mice express a strong resistance to challenge if the adults arising from infection are removed by anthelmintic, or if the infecting larvae are irradiated so that they cannot develop into adult worms. In immune mice challenge larvae become trapped in the mucosa and are eventually destroyed in inflammatory granulomata, suggesting that ADCC responses can act as an important effector mechanism.

Experiments of this type point to two important conclusions: that mucosal larval stages are protectively immunogenic, and that adult worms in some way interfere with the development or expression of immunity. Both conclusions were elegantly confirmed in experiments in which irradiated larvae were used to immunize groups of mice, some of which had previously had adult worms implanted directly into the intestine. When the mice were challenged, those given only irradiated larvae were highly resistant, whereas those given adult

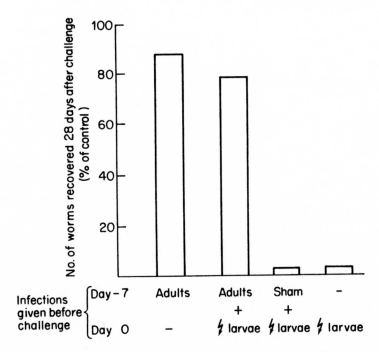

Fig. 7.10   Adult worm-induced modulation of immunity against *Heligmosomoides polygyrus* in mice. Mice were given 100 adults by laparotomy and/or 200 irradiated larvae. After removing worms by anthelmintic treatment on days 40 and 42 the mice were challenged with 200 normal larvae on day 49. (Redrawn from Behnke *et al.*, 1983, *Parasite Immunology*, **5**, 397.)

worms and irradiated larvae were almost as susceptible as the controls. (Fig. 7.10).

The mechanisms of the suppression exerted by adult worms are uncertain. Mice infected with *H. polygyrus* show depressed responses to a number of unrelated antigens including sheep red blood cells (SRBC), influenza virus and other intestinal nematodes. Depression of SRBC responses is associated with impaired antigen presentation by macrophages and with the appearance of T cells that act suppressively. Mice concurrently infected with adult *H. polygyrus* and *T. spiralis* fail to express immunity against the latter and show a minimal mucosal mastocytosis.

These data suggest that adult *H. polygyrus* may interfere in some way with the immune response at a fundamental level, perhaps by secreting immunomodulatory factors that alter T cell responses. Support for this idea has come from analysis of cytokine responses to infection in susceptible and resistant strains of mice. The mesenteric node lymphocytes of susceptible mice produce only moderate amounts of IL-3, IL-4 and IL-9 during the first few weeks, and no IL-10. In contrast, cells of resistant mice secrete large amounts of all four cytokines and sustain secretion of IL-3 and IL-9 for some time, i.e. they make a good Th2 response. This is not to say that there is only one mechanism for effective resistance, indeed different strains of mice appear to express resistance in different ways, as is suggested by the fact that F1 hybrids between SWR and SJL mice respond more quickly than either parental strain (Fig. 7.9).

Trichuris muris is a natural parasite of mice, but is normally found only in relatively small numbers. In most strains of laboratory mice, infection stimulates a strong immunity that removes the worms before they mature and provides complete resistance to reinfection. However, in certain strains, all mice will become heavily infected. It appears from detailed studies of T cell and cytokine responses that immunity requires Th2-mediated responses, but these can be diverted into Th1 responses as worms develop. There is a crucial period during the fourth and fifth weeks of infection when worms are either expelled or they modulate the host's response from Th2 to Th1. Once this has been achieved the mice seem to become permanently tolerant to the parasite, and even if an initial mature infection is removed, a second infection will still develop normally. This susceptibility to the immunomodulatory influence of the parasite is genetically determined by both background and MHC-linked genes.

7.5.2   Biological significance of chronic infections

In biological terms chronic infections, however achieved, undoubtedly favour parasite survival, since they allow prolonged reproductive activity. It is not necessary for all hosts to carry chronic infections, merely that enough should do so to maintain an adequate rate of transmission to susceptible individuals. One of the characteristic features of nematode infections in natural populations, including those in humans, is that the worms are aggregated, i.e. the majority of hosts have few or no parasites, and the majority of worms occur in a few hosts. These heavily and chronically infected individuals maintain parasite transmission and so ensure parasite survival. Many factors influence this aggregation, some are environmental (e.g. the distribution of infective stages), others behavioural (e.g. patterns of working or feeding), but many operate through influences on host immunity. Genetically determined differences in ability to develop resistance are certainly important, and have been thoroughly examined in experimental models. Immunity to GI parasites seems to operate inefficiently in young animals and is very effectively depressed by the physiological changes that accompany pregnancy and lactation, particularly the latter. Reproductively active females therefore provide an important source of infection for their susceptible young, who may then respond ineffectively and become chronically infected. In sheep, parasitized by GI nematodes, for example, parturition and lactation are associated with a marked elevation of faecal egg ouput (the Spring Rise in the Northern hemisphere) as immunity is depressed and as larval nematodes whose development in the host has been inhibited become mature. The net effect is that pasture becomes heavily contaminated with infective larvae, ensuring transmission to newborn lambs.

8

Nematodes that invade tissues

The cuticle as target for effector mechanisms

8.1 Introduction

Exploitation of habitats other than those provided by the intestine of the host is widespread in the Nematoda. Many species that live as adults in the intestine, e.g. *Ascaris*, hookworms, *Nippostrongylus* and *Trichinella*, undergo development in parenteral tissues. Other species are wholly confined to the tissues and have no contact with the intestine. The occupation of such niches within the body requires particular adaptations in reproductive biology, the parasite concerned no longer having direct access to the outside world. One way of solving this problem (seen in *Capillaria hepatica* and *Trichinella*) is the production of stages, eggs or cysts, which remain infective in the tissues after the death of the host or that are infective when the host is eaten. Another (seen in *Dracunculus*) is to break out of the surface of the body in order to liberate larvae. In a major group of tissue-invading nematodes, the Filarioidea, the problem is solved by the involvement of a blood-feeding arthropod intermediate host in the life cycle (Fig. 8.1). The female worms liberate live embryos – microfilaria larvae – which circulate in the blood or accumulate in the skin. The arthropod takes up microfilariae when it feeds, provides an environment in which development to infectivity can occur, and then reintroduces the parasite into the vertebrate host at a subsequent blood meal. For the major filarial infections of man the intermediate hosts are as follows:

Lymphatic filariases: *Wuchereria bancrofti*, *Brugia malayi* – transmitted by species of mosquito.

Onchocerciasis: *Onchocerca volvulus* – transmitted by species of *Simulium*.

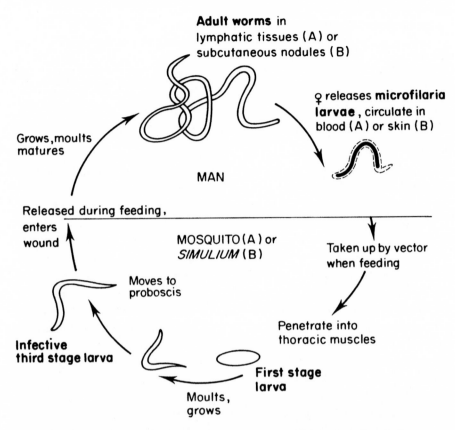

Fig. 8.1 Life cycles of human filarial nematodes. A, *Wuchereria* or *Brugia*; B, *Onchocerca*. The life cycles of the common laboratory models *Brugia pahangi* and *Acanthocheilonema viteae* are similar to A. *A. viteae* adults lie unencapsulated below the skin and their larvae are transmitted by ticks.

However, many other arthropods, including ticks and mites, can also transmit worms of this group.

In several species of filarial worms there is an intricate interrelationship between the behavioural patters of parasite and vector, which results in an optimization of the uptake of microfilarial stages. The latter are not present in peripheral blood throughout the 24-hour cycle but appear for limited periods during the day or during the night. The periodicity of the parasite coincides with the time at which the vector feeds most actively. In *Wuchereria bancrofti*, for example, the microfilaria larvae appear in the peripheral blood for an hour or two each side of midnight, when the mosquito is feeding. For the remainder of the 24 hours the larvae remain within the deep organs of the body, particularly

Table 8.1 *Major filarial infections of humans*

Species	Vector	Adult	Microfilaria
Wuchereria bancrofti	Mosquito	Lymphatics	Blood
Brugia malayi	Mosquito	Lymphatics	Blood
Onchocerca volvulus	*Simulium*	Skin	Skin

the lungs, actively maintaining position against the flow of blood. Movement into and out of peripheral blood is controlled by the physiological rhythms of the host and is reversed if the waking–sleeping pattern of the host is reversed.

Filarial infections are responsible for some of the most important parasitic diseases of man and they are endemic in areas that allow the development of suitable intermediate hosts (Table 8.1). All are long-term, chronic diseases, characterized by morbidity rather than by mortality. They are associated with debilitating pathological changes, particularly in the lymphoid system and in the skin. Swelling and inflammation of the lymph nodes occur commonly with all three species listed above; inflammation and blockage of lymphatics is characteristic of *Wuchereria* and *Brugia*. Changes in the skin and dermal tissues are most pronounced in onchocerciasis and this infection also leads to severe eye lesions and the loss of sight. Both lymphatic species may induce skin and dermal changes in the condition known as elephantiasis. Pathology in onchocerciasis is due entirely to the microfilarial stage; in the lymphatic species it is primarily the adult worms that are harmful. It is known that many of the changes associated with filarial infection are immunopathological in origin and that hypersensitivity reactions are particularly important in their development. It must be remembered, however, that not all human filarial infections are severely pathogenic, a number are quite benign despite their chronicity and give rise to very few clinical symptoms.

The existence of immunopathological changes in filarial disease reflects the close and intimate contact between tissue-dwelling worms and the immunological mechanisms of the host. Parasite antigens are freely available to the host and, unlike gastro-intestinal species, the worms are readily access-ible to effector agents such as antibody, complement and cytotoxic cells. It is therefore possible to think of immunity against such worms in conventional terms, the direct interactions between effectors and the worm playing a major role. It might be thought that, under these conditions, host responses would constitute an effective barrier to successful parasitism, but clearly this is not the case. Filarial nematodes are not only large and successful parasites, infecting an estimated 300 million people, but their life span is considerable, individual worms surviving for many years in the host. *Onchocerca volvulus*, for example, may live for more than 15 years. This long-term survival is the more remarkable when it is apparent that infections do elicit strong immune responses, most characteristically reflected in raised levels of IgE and pronounced eosinophilia.

8.2   Protective immune responses

The chronicity of filarial infections implies that protective responses are absent or weak, that worms evade such responses by reducing their immunogenicity or that they suppress the host. The evidence for protective immunity in humans is essentially circumstantial and comes from two categories of epidemiological observation. The first is that infection levels may plateau or decline with age, the second that in endemic areas there are always individuals in the population who appear to be parasitologically negative, despite immunological evidence for exposure to infection. *In vitro* studies using human cells and sera have also shown that infection generates antibodies capable of mediating cytotoxic responses against parasite stages (see below). Conclusive data that filarial infections elicit protective immune responses can only come from experimental work, but it is difficult to investigate responses to the human species of filaria in this way because of their host specificity. However, related species can be maintained in laboratory hosts and these, together with rodent filaria, provide valuable models of both those situations in which protective responses do develop and those in which immunity is ineffective (Table 8.2). The majority of these infections essentially model lymphatic filariasis, although features of all are generally relevant to analysis of immune responses to onchocerciasis.

Effective immunity may result in the death of adult worms or their loss of fecundity, in clearance of microfilariae from the bloodstream, or prevent the development of infective larvae and juvenile stages so that challenge infections fail to become fully established. Where there is no immunity, hosts are unable to control primary infections or resist challenge infections even when repeatedly infected over long periods of time (Fig. 8.2). Immunity in many cases, and probably all, is clearly T cell-dependent. Nude mice and rats are susceptible to infection with a number of species (*A. viteae, B. pahangi, B. malayi*) to which their normal, immune competent littermates, are resistant, and infections with *B. pahangi* can be established in normally resistant CBA mice when these are deprived of T cells.

Table 8.2 *Major laboratory models used to study immunity to filarial infections*

Acanthocheilonema viteae	Jird, Hamster, (Mouse/Rat)
Brugia malayi	Jird, Ferret, (Mouse)
Brugia pahangi	Cat, Jird, Rat, (Mouse)
Dirofilaria immitis	Dog
Litomosoides carinii	Cotton rat, Jird, Rat
Onchocerca spp	Cattle, (Mouse)

Notes:
(Host) signifies use for establishment of infection by transfer of adult worms or injection of microfilaria. Immune-deficient animals may allow direct infection by L3 larvae.

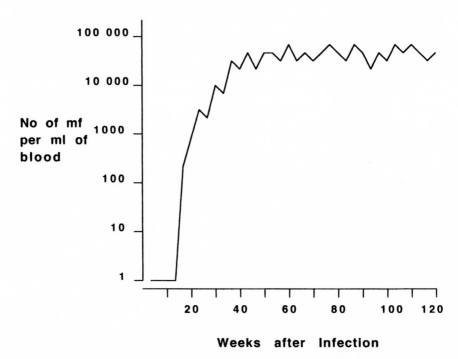

Fig. 8.2   *Brugia pahangi* in cats. Microfilaraemia (number of microfilaria, mf, per ml of blood) in an individual given 25–50 infective L3 larvae weekly. (Redrawn from Denham *et al.*, 1992, *Parasitology*, **104**, 415.)

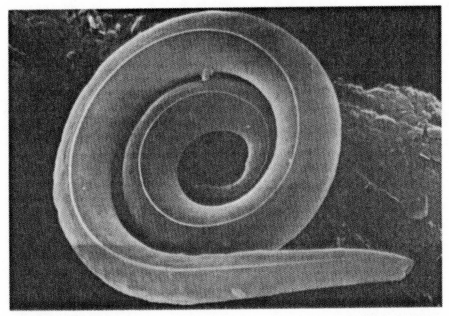

Fig. 8.3   Unsheathed microfilaria, showing cuticle and oral opening. (Photograph by courtesy of Dr S.L. Croft.)

Many studies of immunity have concentrated upon responses directed against microfilarial stages, as these are easily available and amenable to *in vitro* manipulation. Analyses of responses against the infective and adult stages present more difficult technical problems.

8.3   Responses against microfilaria larvae

It has been shown many times, in both human and experimental hosts, that infection leads to production of antibodies directed against the microfilarial sheath, or cuticle in unsheathed forms (Fig. 8.3). These antibodies can be detected by immunofluorescence or by their ability to mediate attachment of cells. Two observations in particular have focused attention on the role of such antibodies in host protection. In many cases, antibodies specific for surface antigens do not appear until after the host becomes amicrofilaraemic, i.e. when larvae are no longer detectable in the blood. Anti-microfilarial antibodies therefore correlate much better with resistance than do those against adult worms. Secondly, antibody-mediated cell adherence can lead to cytotoxicity and result in the death of the larvae. By implication, the antigens

Table 8.3 In vitro *killing of microfilaria larvae by ADCC*

Species	Antibody	Cell type
A. viteae	IgM, IgE	E/M/N
B. malayi	IgG	E/M/P
B. pahangi	IgG	E/N
D. immitis	IgM	N
L. carinii	IgE	E/M
O. volvulus	IgE	E/N

Notes:
E, eosinophil; M, macrophage; N, neutrophil; P, platelet.
Source: (Modified from Piessens *et al.*, 1990, in Wyler,
Modern Parasite Biology, W.H. Freeman.)

that elicit these responses must be present at the surface of the larvae, and the effector mechanisms that destroy the larvae must do so initially by damaging their outer layers.

A variety of antibody isotypes and effector cells appear to be capable of killing microfilariae (Table 8.3). The essential role of cells in killing micro-filariae was elegantly demonstrated by experiments in which larvae of *A. viteae* were placed within micropore chambers, that were then implanted into immune hamsters. When chambers with a pore size of 0.3 μm were used, no cells were able to enter the chambers and the larvae survived for at least 3 weeks. When chambers with a pore size of 3 or 5 μm were used, cells were able to enter and the larvae were killed within 24 hours. The same result was obtained when microfilariae were pre-incubated in the IgM fraction of immune serum before being placed in 3 mm pore size chambers and implanted into normal hamsters.

This *in vitro* evidence for an interaction between IgM and white cells in larval killing is supported by *in vivo* data gained by using a mutant strain of mice (CBA/n) that is genetically incapable of making IgM responses. In these mice microfilaraemia established by transplantation of adult female *A. viteae* remains stable for long periods (> 180 days), whereas in normal CBA the numbers of larvae fall rapidly. There is no difference between the two strains in adult worm survival, showing that the difference in duration of micro-filaraemia was due to an immunity that acted specifically against this life cycle stage. Similar data have been obtained from studies using strains of mice that are immunologically normal, but which show quite different capacities to

**Microfilaria
/10µl blood**

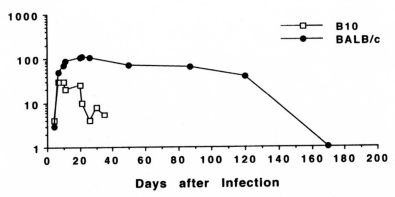

Fig. 8.4   Time course of microfilaraemia in C57BL/10 (B10) and BALB/c mice after subcutaneous implantation of five female *Acanthocheilonema viteae*. (Redrawn from Storey *et al.*, 1987, *Acta Tropica*, **44**, 43.)

control microfilaraemia. In BALB/c mice, microfilaraemia is prolonged for more than 200 days, whereas in C57BL/10 mice microfilaria are cleared within 50–75 days (Fig. 8.4). This difference is genetically determined, but independent of MHC-linked genes. Although the two strains do show differences in ability to make anti-parasite IgM, it is clear that other factors are also involved in their different responses to infection. Antibodies from BALB/c mice appear to have a lower affinity for microfilarial antigens, i.e. they bind less well and therefore are less efficient in ADCC. Transfer of immune sera from C57BL/10 donors into microfilaraemic BALB/c mice causes a rapid fall in microfilaraemia, but this is then slowly reversed (Fig. 8.5), suggesting that BALB/c mice may also be defective in a mechanism necessary for subsequent microfilarial elimination.

Detailed *in vitro* studies have also been made of ADCC killing of *A. viteae* microfilariae using the rat model. The antibody isotype primarily involved here is IgE, which acts as an opsonizing antibody, arming cells for interaction with the surface antigens of the parasite. In the presence of immune serum, cell adherence occurs rapidly. Larvae become heavily coated with cells by 16 hours and many appear dead at this time. The first cell type to adhere is the eosinophil, which makes close contact with the cuticle and degranulates onto the surface. The layer of material released from the cells then lifts away and macrophages adhere in large numbers, actively phagocytosing dead cells. After macrophages adhere they spread over the cuticular surface and release

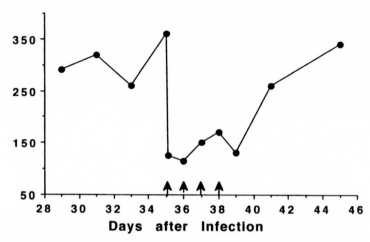

Fig. 8.5   Effect of repeated transfer of immune serum on the microfilaraemia of BALB/c mice with a transplanted *Acanthocheilonema viteae* infection. Arrows indicate days on which serum was injected intravenously. (Redrawn from Storey *et al.*, *International Journal for Parasitology*, 1989, **19**, 723.)

lysosomal material. At this stage the cuticle shows visible damage and internal tissues begin to lyse.

An interesting and important aspect of this study was that antibodies capable of mediating ADCC appeared only after the rats became amicrofilaraemic. This is in line with evidence from infections in humans, and supports the idea that anti-microfilarial responses may be an important component of human protective responses to the lymphatic filariae. One puzzle is why it takes a comparatively long time for these antibodies to appear. It may be that large numbers of circulating larvae adsorb out serum antibodies, or it may be a consequence of the immunosuppression associated with infection (see below). It is also possible that microfilaria evade recognition by changing their surface antigens as they develop or disguise them by acquiring host molecules. Labelling studies show that the surface molecules of newly released larvae differ from those of older larvae, and there is evidence in some species that larvae acquire host serum albumin on their surface.

8.4   Responses against infective larvae and adults

Epidemiological studies in endemic areas show that virtually all members of a population will be exposed to bites of infected vectors, yet a proportion never develop patent infections or clinical signs of infection even though they may be immunologically positive, i.e. have been exposed to the parasite. This may reflect a resistance that is effective against the infective larvae or later pre-adult stages. Other parasitologically negative individuals do show clinical signs of infection and may suffer, for example, from recurrent fevers and inflammation of the lymph nodes. This again may indicate that there is a resistance against the initial stages of infection or show that there is immunity against the adult worms that prevents reproduction and release of microfilariae.

Although direct evidence that immunity operates against the infective or pre-adult stages is difficult to obtain *in vivo,* there are data showing that it does occur. For example, the L3 of *B. pahangi* are known to be killed by immune-mediated mechanisms. Cats that have become amicrofilaraemic after repeated reinfection with *B. pahangi* are resistant to challenge, and L3 stages are killed before they enter the host's lymphatics. The immunogenicity of the infective stage is confirmed by the fact that irradiated live L3 have been used successfully to vaccinate against *B. malayi* in monkeys, *B. pahangi* in cats and birds, *Dirofilaria immitis* in dogs and *L. carinii* in rats. In addition, studies in human populations infected with *W. bancrofti* have shown the presence of antibodies against L3 surface antigens in adults who, although repeatedly infected, show stable worm burdens.

Circumstantial evidence in humans suggests that immunity can also operate against the adult worms, and this is reinforced by experimental data. However, the way in which immunity acts varies considerably between different laboratory models. In some, e.g. *A. viteae* in the rat or mouse, adult worms die long before microfilariae are cleared from the blood. In others, the host becomes amicrofilaraemic even though the adult worms remain alive. This occurs during infections with *A. viteae* in hamsters, where it has been shown that the disappearance of microfilariae from the circulation (latency of infection) is in part due to a failure of the female worms to release larvae. Latent worms resume microfilarial release when the host is immune suppressed, or when they are transferred from latent into naive hosts. Antibodies against adult worm antigens are readily detectable in infected hosts, but they do not necessarily correlate with resistance to infection. Indeed many antibodies are targeted against internal antigens that would not become available to the host until after the death of the worms and could not otherwise act as immunogens.

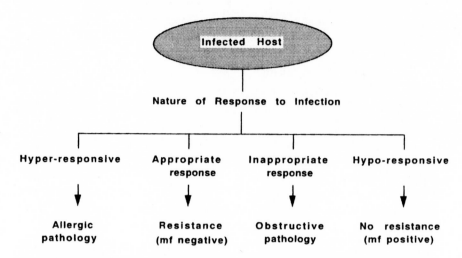

Fig. 8.6   Diversity of host response in lymphatic filariasis. (Redrawn from King & Nutman, 1991, *Immunoparasitology Today*, **12**, A54.)

An important question, that is still not resolved, is whether there is complete immunity to filarial infections in humans, i.e. whether endemic normals are wholly immune and parasite-free. Studies in Papua New Guinea showed that all individuals in an area endemic for *W. bancrofti,* even those who were amicrofilaraemic and asymptomatic, showed serological evidence of the presence of adult worms, being positive for a circulating 200 kDa phosphorylcholine-containing antigen and for parasite-specific IgG4 antibodies. It is possible, therefore, that in filarial infections, as in schistosomiasis, there is a degree of concomitant immunity, immunity against incoming L3 preventing futher infection but not affecting the survival of existing adult worms.

8.5   Immunomodulation and immunopathology in filariasis

The spectrum of infection and pathology seen in populations exposed to lymphatic filarial infections is very broad (Fig. 8.6). At one extreme are the individuals who are parasitologically negative but immunologically positive (endemic normals), at the other are those who are heavily infected, as judged by the microfilarial loads, but immunologically hyporesponsive (partially tolerant) to filarial antigens. Endemic normals and heavily infected individuals may show little or no pathology, whereas others, who are amicrofilaraemic, may show the severe pathology associated with lymphatic inflammation and obstruction (elephantiasis) and have strong cellular and antibody responses. A fourth category appear to be hyperresponsive to infection and may develop abnormal symptoms such as tropical pulmonary eosinophilia (TPE). Current interpretations of this spectrum of conditions is focused on the concept that the outcome of infection is determined at the T cell level by interactions between genetically and environmentally determined characteristics of the host and parasitological factors determined by the level, frequency and duration of exposure to infection.

Two characteristic responses to filarial infection (shared with many helminths) are eosinophilia and elevated levels of serum IgE. It seems clear that these components of the immune system may help to provide protection (e.g. in ADCC) but can also contribute to immunopathology. For example, individuals with TPE have high levels of IgE and circulating eosinophils. In general, those with lymphatic pathology and who are amicrofilaraemic have higher levels of IgE and IgG (particularly IgG3) than those who are asymptomatic but microfilaraemic (Fig. 8.7a). The latter tend to have higher levels of the IgG4 isotype (up to 95% of total filarial-specific antibody) a finding of some significance, because IgG4 antibodies can competitively block IgE-dependent allergic responses by binding to antigens in tissue fluids before they come into contact with IgE bound to mast cells or basophils (as in Schistosomes, p. 117). Both IgE and IgG4 responses, as well as eosinophilia, are dependent upon release of cytokines associated with the Th2 subset of helper cells, such as IL-4 and IL-5. Microfilaraemic individuals also show much reduced T cell proliferative responses to filarial antigens, whereas endemic normals and elephantiasis cases often show high responsiveness (Fig. 8.7b). This complex of response patterns can perhaps be explained by parasite-associated interference with T cell activation and regulation. Lack of responsiveness to antigens may reflect an interference with Th1 function, through altered antigen presentation or IL-2 responses, or increased release of cytokines such as IL-10 that turn off Th1 responses. Individuals with active infections certainly do show suppressed production of the Th1 cytokine IFN-γ. (Fig. 8.7c). Depression of Th1 activity may correlate with an increase in Th2 function, leading to production of IgG4, IgE and eosinophilia. Clearly, the different levels of IgG4 and IgE seen in various groups must result from a more subtle influence upon Th2 function, since both isotypes are Th2-dependent. An important factor in generating particular patterns of responses to infection may be prenatal exposure to parasite antigens.

The various pathological changes that accompany lymphatic filarial infections can be seen as consequences of these patterns of immune response and

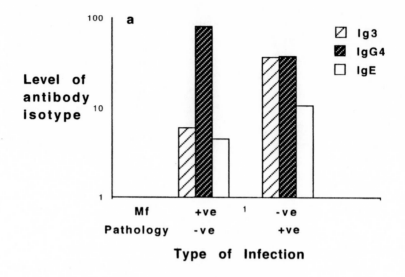

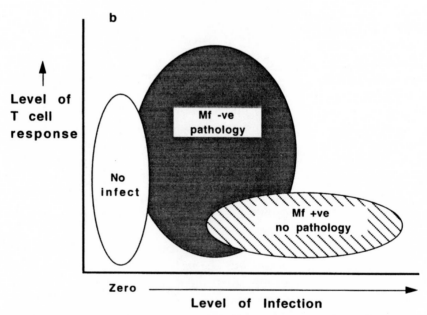

Fig. 8.7 Variations in immune and pathologic responses to lymphatic filariasis (*Brugia malayi*) in individuals showing positive microfilaraemia but no pathology, or negative microfilaraemia and obstructive pathology.
(a) Levels of parasite-specific IgG3, IgG4 and IgE (μg ml^{-1})
(b) Levels of T cell responsiveness (schematic). (No infect, individuals living in the endemic area but showing no parasitological evidence of infection, therefore possibly immune.)

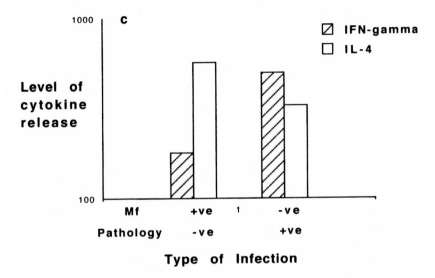

Fig. 8.7 (*cont.*)
(c) Levels of interferon-gamma (IFN-gamma ng ml^{-1}) and IL-4 (pg ml^{-1} × 100)
released from peripheral blood mononuclear cells stimulated with filarial antigen.
(Redrawn from Maizels *et al.*, 1995, *Parasitology Today*, **11**, 50.)

the stages of infection that act as targets. Individuals who fail to prevent repeated reinfection, but respond strongly to the juvenile, adult and microfilarial stages, are likely to develop the characteristic lymphatic pathology (Fig. 8.8), whereas those who have an effective anti-L3 immunity may escape this, particularly if they maintain levels of microfilaraemia that suppress responsiveness. In *Onchocerca* infections pathology is associated almost exclusively with the microfilaria, which initially provoke hypersensitivity reactions in the skin leading to intense dermatitis and itching. The skin then becomes infiltrated with mast cells, eosinophils, lymphocytes, plasma cells and macrophages. At later stages of infection there is a relative unresponsiveness to parasite antigen, probably reflecting interference with T cell activity, but degenerative changes continue in the skin. Invasion of the eye by microfilaria can also produce chronic inflammatory changes, leading eventually to blindness.

8.6   Antigens of filarial worms

A complete understanding of the complex interactions of filarial infections with the host immune response obviously requires detailed knowledge of the antigens and other immunologically active molecules released from the worms.

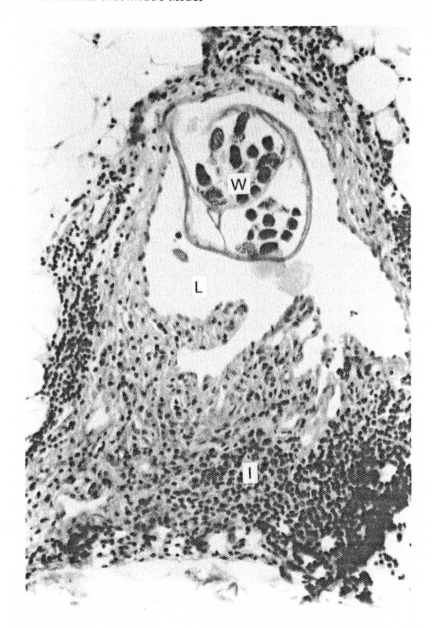

Fig. 8.8   *Brugia pahangi*. Section showing adult worm (W) in the inflamed lymphatic of an experimentally infected ferret. I, inflammatory infiltrate; L, lumen of lymphatic. (Photograph by courtesy of Dr R.B. Crandall.)

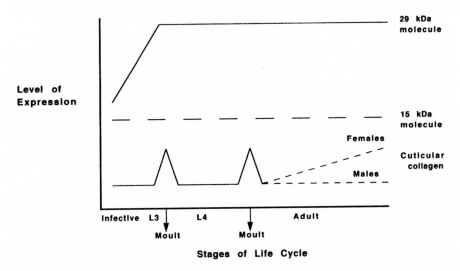

Fig. 8.9   Time course of synthesis of surface-labelled proteins of *Brugia malayi* in the mammalian host. Expression of the 15 kDa molecule is essentially constant, whereas the 29 kDa molecule is expressed only after entry into the mammal. There is little synthesis of collagen between moults or in the adult male worms; the increase in the adult females reflects synthesis of microfilarial cuticle. (Redrawn from Selkirk, 1991, in *Parasitic Nematodes – Antigens, Membranes, Genes*, ed. M.W. Kennedy, p. 27, Taylor & Francis.)

Although many have been identified, and a number are now cloned, in most cases their biological functions are largely unknown. However, given that the filarial cuticle is known to be metabolically active and to mediate a number of physiological functions, including nutrient uptake, it is likely that some of the molecules identified have transport or secretory roles. More recently it has become clear that some of the molecules may also protect against host defences.

Surface antigens have been identified on the developmental stages of a number of species, including *W. bancrofti*, *B. malayi*, *B. pahangi*, *O. volvulus* and *A. viteae*. Many of these are also released from the worms. Although there is some stage- and species-specificity there is also a great deal of cross-reactivity. One of the major surface antigens of all lymphatic species is the glycoprotein gp29. This is a homologue of the enzyme glutathione peroxidase. It may therefore function as an anti-oxidant enzyme and thus help to protect the surface against potentially damaging mediators released from host white cells. In contrast to other molecules (e.g. the 15 kDa surface antigen) gp29 is not fully expressed until after entry into the mammalian host and is difficult to detect in the mosquito stage. It may therefore represent an adaptive response allowing survival in the new host environment (Fig. 8.9).

The high degree of cross-reactivity between antigens from different filarial worms is striking. This may not be surprising if the molecules concerned have essential metabolic functions. Another interpretation is that cross-reactivity may reflect evolutionary pressures on species to conserve molecules that are effective in down-regulating protective immunity in the host. Many antigens contain the phosphorylcholine determinant, which may divert antibody responses away from functional antigens, but which may also help to suppress lymphocyte responsiveness.

9

Ectoparasitic arthropods
Immunity at the surface of the body

9.1   Introduction

Terrestrial vertebrates act as hosts for a variety of ectoparasitic arthropods. Infestation with such parasites may give rise to pathological symptoms, through damage to dermal tissues or systemic effects, but of far greater significance is the transmission of infectious organisms, from viruses to helminths, for which arthropods act as vectors. As a way of life, ectoparasitism is harder to categorize than endoparasitism. In the latter the host provides the total environment, contact is intimate and prolonged, and the parasite exhibits complete metabolic dependency. In the former the parasite may make only intermittent contact for the purposes of feeding, and spend long periods away from the host. As with all biological phenomena, however, there is a wide spectrum of associations covered by the term ectoparasitism and this is reflected both in the duration and the intimacy of host contact. In mammals, for example, the extremes can be illustrated by the scabies mite (*Sarcoptes scabei*), which spends its whole life upon the host and burrows into the skin, and by the mosquito, in which only the female is parasitic, making sporadic host contact in order to feed briefly before egg laying.

9.2   Responses to feeding

With all ectoparasites, the point of host contact is the skin and it is through this organ that protective responses are initiated. If the parasite is accessible, protection may be achieved through the entirely non-specific mechanisms of scratching or grooming. Immune responses, particularly immediate hyper-sensitivity reactions, may enhance this non-specific protection by making the host more aware of the presence of the parasite. Where the parasite is less accessible, less easily dislodged, or where the host is less able to groom, then immune responses with a direct, anti-parasite effect may play an important role in protection.

The skin is well equipped to make protective responses, its extensive vascularization providing ready accesss for humoral and cellular effectors. In addition, the resident amine-containing cells form a sentinel population capable of rapid reaction to tissue damage and, by their degranulation, allow-ing enhanced infiltration of other cell populations. There is a characteristic population of dendritic, antigen-presenting cells, the Langerhans cells, which are ideally situated for processing antigens presented during the feeding of the parasite. All ectoparasites introduce saliva into the wound made by their mouthparts. In many cases this contains anticoagulant factors, which prevent blood from clotting, histolytic enzymes, vasoactive amines and toxins. The proteins present in saliva are potent immunogens and elicit strong immune responses, frequently hypersensitive in nature. The effect of these responses upon the parasite is of course related to the duration of feeding. If this is less than the time necessary for the development of the response then the parasite feeds and escapes unscathed. With longer feeding times the response gener-ated has the opportunity to affect the parasite. The nature of skin-based pro-tective responses has been investigated most thoroughly in host–parasite associations involving ticks.

9.3   Immune responses to ticks

Ticks belong to a large group of the arachnid arthropods, the Acarina. They are distinguished from the closely related mites by their larger size and by their host relationships. All are parasitic and all are blood feeders throughout their development. Many transmit serious pathogens to man and to domestic animals and are major pests, particularly in countries with warm climates (Table 9.1).

Ticks show a variety of host-contact patterns during their life cycles. In some species each developmental stage feeds upon the same host individual, in others two or three individuals are used, with the ticks leaving the host when replete in order to moult. In three-host ticks a different individual is used by each stage in the cycle, i.e. larva, nymph and adult, and one blood meal is taken on each host. (These patterns characterize the hard (ixodiid) ticks, soft

Table 9.1 *Pathogens and diseases transmitted to man and domestic animals by ticks*

Genus	Pathogen	Disease
Soft ticks		
Ornithodorus	Spirochaetes	Relapsing fever
Hard ticks		
Amblyomma	Rickettsias	Spotted fever
	Bacteria	Tularaemia
Boophilus	Protozoa (*Babesia*)	Piroplasmosis
Dermacentor	Viruses	Encephalomyelitis
	Rickettsias	Spotted fever
	Bacteria	Tularaemia
	Protozoa (*Babesia*)	Piroplasmosis
Haemaphysalis	Rickettsias	Spotted fever
	Bacteria	Tularaemia
	Protozoa (*Babesia*)	Piroplasmosis
Hyalomma	Rickettsias	Rickettsiosis
	Protozoa (*Theileria*)	East Coast fever
Ixodes	Viruses	Encephalomyelitis
	Rickettsias	Tick typhus
	Bacteria	Lyme disease
	Protozoa (*Babesia*)	Piroplasmosis
Rhipicephalus	Viruses	Encephalomyelitis
	Spirochaetes	Tick typhus
	Protozoa (*Babesia, Theileria*)	Piroplasmosis, East Coast fever

(argasid) ticks feed more frequently.) The feeding of hard ticks is a lengthy process and may take several days. The mouthparts are adapted for cutting into the skin and for taking up blood. Feeding is preceded by production of saliva and of a cement-like material that serves to hold the mouthparts firmly in the skin (Fig. 9.1). The antigenic material contained within these secretions persists in the skin for several days after feeding and can be identified, by fluorescent labelling, on the surface of the Langerhans cells. Even on naive hosts tick bites lead to pronounced inflammatory responses, but these do not prevent engorgement. In the immune host the reaction to biting is rapid and may prevent feeding completely. Failure to feed quickly causes the death of the tick, from the combined effects of starvation and desiccation.

Immunity to ixodiid ticks was first described as long ago as 1939 by Trager, who found that larval stages of *Dermacentor variabilis* were completely unable to feed upon guinea-pigs that had been exposed to repeated experimental infesta-

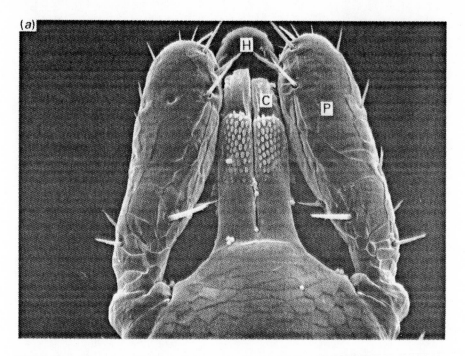

Fig. 9.1(a)   Dorsal view of mouthparts of an *Ixodes* tick. C, chelicera, used to pierce skin; H, hypostome, used to anchor mouthparts in skin; P, palp, sensory function. (Photograph by courtesy of Professor Dr H. Melhorn.)

tions. It has also been known for many years, initially from empirical field observations, that there is acquired resistance to ticks in cattle, some individuals and some breeds developing greater immunity than others. In recent years guinea-pig and mouse models of tick immunity have been intensively studied. The kinetics of the response have been described in detail and the underlying mechanisms have been very fully analysed. Immunity is easily assessed in this system and is usually measured from the percentage of ticks that successfully complete their feeding and become fully engorged (Table 9.2).

9.4   Analysis of immunity

Many studies, using a variety of ixodiid ticks, have shown that an effective and long-lasting immunity develops extremely rapidly, being apparent within a week of initial infection. On immune hosts ticks fail to engorge, show impaired moulting, and may die from desiccation within 24 hours. The immune response to infestation is remarkably sensitive. Experiments have

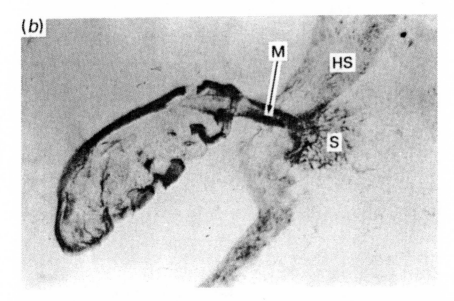

Fig. 9.1(b)   Section through larval *Boophilus* feeding on a bovine, showing extent of secretion (esterase-positive) released into the skin. M, mouthparts; HS, host skin; S, secretion. (Photograph by courtesy of Dr P. Willadsen and P. Tracey-Patte.)

shown, for example, that guinea-pigs exposed to one mature female tick for more than 48 hours develop nearly complete immunity against a subsequent challenge with 200 larvae. As this indicates, immunity is not stage-specific: all stages in the life cycle can immunize against larvae, and larvae immunize against nymphs.

   Immunity is not a local phenomenon and once elicited will operate effectively at sites remote from the original feeding area. Transfer of immunity can be achieved using either immune serum or immune lymphocytes and very high levels of protection can be conferred on recipients. In guinea-pigs, for example, transfer of 1 ml of immune serum, or 2×10^8 immune cells, gave 90% protection against challenge with *Rhipicephalus appendiculatus*. The degree of immunity that can be transferred varies between systems, depending on the species of tick and the breed of host.

9.4.1   Cellular and antibody basis of immunity

When ticks feed on a naive host the first cells to infiltrate the attachment site are neutrophils, followed at a later stage by basophils, mast cells and eosinophils. The relative importance of basophils and mast cells as amine-

Table 9.2 *Effect of immunity established by initial infestation upon development of challenges with tick larvae in guinea-pigs*

	Amblyomma		Rhipicephalus	
	% recovery of ticks	Weight of ticks (mg)	% recovery of ticks	Weight of ticks (mg)
Larvae on control hosts	80	0.95	73	0.28
Larvae on immune hosts	34	0.58	28	0.20
% rejection of larvae	58	–	62	–
% reduction in weight	–	39	–	29

Source: (Data from Brown & Askenase, 1981, *Journal of Immunology,* **127**, 2163.)

containing cells in the response varies according to the host. Guinea-pigs, rabbits and cattle are basophil-rich animals, in mice mast cells predominate. Tick feeding may cause some degranulation and mediator release from these cells even during a primary infestation. The antigens released from the feeding ticks are processed by the Langerhans cells and presented to T cells, which initiate inflammatory and antibody responses. IgG1 and IgE are produced, both of which can bind to the amine-containing cells; antigen–antibody complexes also form and these may activate complement. The time-scale of this reaction to an initial bite is too slow to seriously inconvenience the ticks, which can complete their feeding normally. However, in an immune host with primed T cells and pre-formed antibody the response is very rapid. In immune cattle and guinea-pigs, tick feeding triggers a massive accumulation of basophils at the dermal–epidermal junction and a smaller, but still significant infiltrate of eosinophils. The number of these cells circulating in peripheral blood also rises. The tissue basophils degranulate in response to antigens secreted by the tick, and their vasoactive mediators cause an oedematous reaction to develop at the attachment site (Fig. 9.2). This interferes with feeding, an effect that is reinforced by the histamine and 5-HT released. Both of these inhibit feeding and production of saliva, and histamine causes the tick to detach. These effects may be the indirect result of changes in the local environment, but it is also possible that mediators exert a direct effect following uptake by the tick, certainly basophils have been seen in the intestines of ticks fed on immune hosts. It was recognized some years ago that the reaction initiated by ticks feeding on an immune host has the characteristics of a cutaneous basophil hypersensitivity response, and it is clear that basophils play a central role. Host resistance can be blocked by

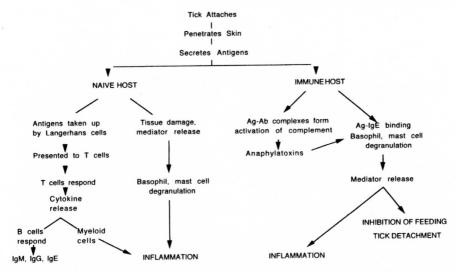

Fig. 9.2   Summary of immune and inflammatory responses induced by ticks feeding on naive or immune hosts. (Modified from Kaufman, 1989, *Parasitology Today*, **5**, 47.)

treatment with a specific anti-basophil antiserum or by injection of anti-hist-amines, and is enhanced by injection of histamine directly into feeding sites. Ticks, however, may gain some protection from a histamine-blocking agent secreted in their saliva.

The importance of amine-containing cells in resistance to ticks has been reinforced by work in mouse models where mast cells rather than basophils are involved (Fig. 9.3). Experiments have shown that mast cell-deficient W/W^v mice have a significantly decreased immunity to the tick *Haemaphysalis longicornis*, but immunity can be restored by reconstituting the mice with bone marrow from normal mice, which restores the skin mast cell response. Even more convincing was the demonstration that protection could be given locally by injection of cultured mast cells into a site challenged by larvae (Table 9.3). When areas of skin grafted from normal donors onto W/W^v mice were used as tick feeding sites resistance was expressed, but when the reverse grafting was carried out (W/W^v to normals) there was no resistance. Experiments involving passive transfer of serum from immune animals showed that the role of mast cells in this response was primarily IgE dependent.

Although amine-containing cells are the most important cell type involved, there is evidence that the eosinophils that accumulate at feeding sites in immune animals can also act as effector cells. Treatment of immune guinea-pigs with an anti-eosinophil serum reduced eosinophil numbers by some 80%

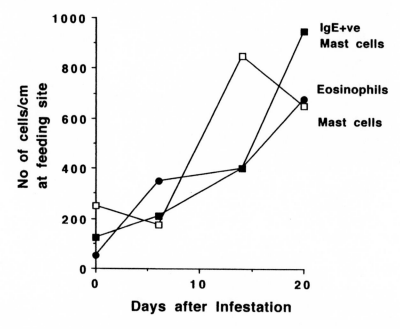

Fig. 9.3   Numbers of mast cells, IgE + ve mast cells and eosinophils appearing at the feeding site in mice given two infestations (days 0 and 14) of the larval tick *Haemaphysalis longicornis*. (Data from Ushio *et al.*, 1993, *Parasite Immunology*, **15**, 209.)

and halved the degree of resistance, but had little effect on numbers of basophils. Eosinophils release a variety of mediators capable of damaging parasite tissues, and their cytotoxic efficiency is increased in the presence of factors (e.g. ECF-A tetrapeptides) derived from basophils and mast cells. In addition, there is evidence that eosinophil peroxidases may have greater activity when bound to released basophil granules.

9.4.2   Tick antigens and immunity

Under natural conditions of tick infestation the immune response of the host is initiated by, and directed against, antigens released in the saliva. Saliva is a complex mixture and several antigens have been identified. It is likely that, during feeding, the host responds to many of these, but some are likely to be more important than others. For example, in the case of *Dermacentor andersoni*, serum from immune rabbits recognizes a large number of antigens with MW ranging from about 30 to 300 kDa. However, certain antigens (e.g. MW

Table 9.3 *Restoration of resistance to infestation by* Haemaphysalis longicornis *in mast cell-deficient W/W^v mice injected with cultured mast cells*

Group of mice (treatment)	% recovery of ticks	Weight of ticks (mg)	No. of mast cells at site of infestation
Normal + medium	56	0.487	495
W/W^v control*	91	0.625	1
W/W^v + medium	84	0.635	1
W/W^v + mast cells	60	0.534	236

Notes:
Mice were given three infestations, two on the right flank, the third on the left flank. Cultured mast cells were injected at the site of the third infestation.
% recovery: recovery from third infestation except in controls*, which were given only a single infestation.
Source: (Data from Matsuda *et al.*, 1987, *Journal of Parasitology*, **73**, 155.)

31 kDa) are secreted early in the feeding cycle but not later, when responses are well under way, and may therefore play a major role in triggering protective responses in immune hosts.

Several early experiments showed that cattle and guinea-pigs could be immunized against tick infestation by injection of crude tick homogenate as a source of antigens. Immunization was effective with or without adjuvant and could give substantial protection, e.g. reducing the numbers of feeding ticks by 65–80%, and decreasing egg production by the survivors. Similar levels of protection could be achieved by vaccinating with antigens prepared only from the midgut of the tick (Table 9.4). The characteristics of this induced immunity, especially its effectiveness against adults rather than larvae, and the degree of protection conferred, suggested the involvement of mechanisms different from those operating in naturally acquired immunity. Adults feeding on vaccinated ticks showed severe damage to their gut cells, damage that could be reproduced by feeding ticks directly on serum or immunoglobulins from vaccinated cattle; complement was not necessary. These results led to the concept of inducing protection by immunizing the host, not against salivary antigens released during the feeding process, but against antigens associated with the intestinal epithelial cells – concealed antigens to which the host is not normally exposed (see Chapter 10).

In cattle, ticks cause considerable damage in their own right, but are equally, if not more, important as vectors of diseases, particularly protozoans such as *Babesia* and *Theileria*. There is a fascinating interplay between the three compo-

Table 9.4 *Immunity to* Boophilus microplus *in cattle vaccinated with midgut antigens in Freund's incomplete adjuvant and exposed to experimental infestation*

Group	No of ticks recovered		No of eggs per normal tick
	Total	Normal	
Control	12 780	11 430	1314
Vaccinated	3321	2088	199
Protection	75%	82%	85%

Note:
Normal = fully engorged ticks.
Source: (Data from Opdebeeck *et al.*, 1988, *Parasite Immunology*, **10**, 405.)

nents of this complex interaction – host, parasite and vector. Breeds of cattle show striking genetically determined variation in their ability to develop immunity against ticks, some, notably the *Bos indicus* breeds (Brahman, Zebu) being considerably more resistant than *Bos taurus* (the common European breeds). Resistance to ticks therefore helps to confer resistance to the protozoans they transmit. However, protozoans such as *Babesia* exert powerful immunosuppressive influences on the host, which in consequence is likely to become more heavily infested with ticks and more frequently reinfected with the parasite. In this way, of course, the protozoan optimizes its own survival and dispersal. Ticks also exert immunomodulatory effects on their hosts, which may compound the immune suppression.

10

Immunological control of parasitic infections

10.1   Introduction

The incentives to devise safe, effective vaccines against parasitic infections remain as great as ever. In the case of human parasites, the difficulties and costs associated with control programmes based on improvements in socioeconomic conditions, sanitation, hygiene or vector control make these strategies difficult to implement successfully on a large scale. The chronic nature of many infections acquired early in life, e.g. malaria, schistosomiasis, filariasis and intestinal nematodes, and the growing evidence that these impair physical and educational development, place a high premium on protection of infants and children. Although chemotherapy has been used to great effect, the widespread occurrence of drug resistance in protozoan and helminth parasites, both human and veterinary, is causing great concern. All of these difficulties strengthen the case for the development of vaccines that can confer long-lasting immunity against the debilitating effects of parasitic infections. In the past, realization of this objective in practical terms has had limited success, and for many years the irradiation-attenuated larval vaccine against lungworm in cattle (DICTOL) was the only vaccine marketed commercially on any scale. Many factors contributed to the failure to develop equally successful vaccines against other parasites. These included:

- lack of knowledge of the mechanisms involved in protective immune responses;
- lack of knowledge of the antigens that elicit protective immunity;

- inability to produce antigens in sufficient quantity;
- failure to vaccinate effectively under field conditions;
- failure to satisfy commercial and user requirements.

10.2   Vaccination against lungworms in cattle

It is instructive to consider the success of the lungworm vaccine in the light of these difficulties. *Dictyocaulus viviparus* is a strongyle nematode that can give rise to epidemics of severe and sometimes fatal respiratory disease ('husk'), particularly in areas where mild, wet climates favour the development on pasture of the infective L3 stages. Infection is acquired when cattle ingest the infective larvae whilst grazing on contaminated pasture. The larvae then migrate from the intestine to complete their development in the lungs (Fig. 10.1). Pathology

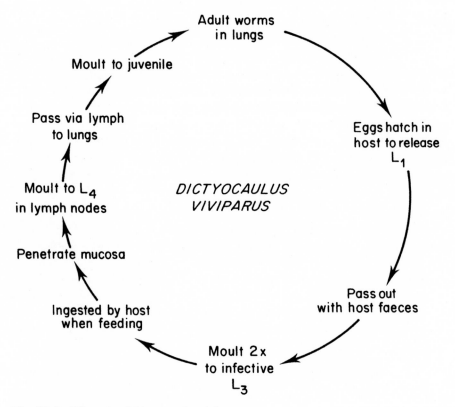

Fig. 10.1   Life cycle of *Dictyocaulus viviparus*.

Table 10.1(a) *Immunity to challenge with* Dictyocaulus
viviparus *in guinea-pigs immunized by infection with 2 ×
5000 40 krad irradiated larvae*

Day after challenge	No. of larvae recovered from lungs after challenge with normal larvae	
	Controls	Immunized
2	140	37
4	69	9
6	136	0
8	375	0

Table 10.1(b) *Comparison of immunization with normal or irradiated larvae*

Day after challenge	Mean larval recovery		
	Controls (no immunization)	Immunized	
		Irradiated larvae	Normal larvae
10	331	0.4	1.0

Source: (Data from Poynter *et al.*, 1960, *Veterinary Record*, **72**, 1078.)

is caused by worms living in the bronchioles, where they induce severe inflammatory responses that can block the airways. Animals surviving the disease are
highly resistant to reinfection. Development of the vaccine was facilitated by
the availability of an appropriate laboratory model. *D. viviparus* can infect
guinea-pigs, and the use of these hosts enabled detailed studies to be made of
the immunogenic and pathogenic stages in the life cycle of the parasite. It was
discovered that, whereas pathology was caused by the adults in the lung, immunity was generated by the larvae during their development to the L4 stage.
Irradiation of L3 larvae at 40 krad allowed the immunogenic stages of the cycle
to be completed but resulted in the death of later stages, thus preventing the
development of disease (Table 10.1a, b). Subsequent experiments then showed
that irradiated larvae gave good protection against disease in calves.

Commercial scaling up of the vaccine was made possible by the ready availability of larvae from the faeces of infected calves. The vaccine is given orally
to calves before they are put out on pasture, using two doses of 1000 irradiated larvae at an interval of 3 to 4 weeks. It does not have to be sterile and can
be administered directly by the farmer. Levels of protection exceed 90%. In

the absence of reinfection immunity wanes, but low levels of infection maintain an effective resistance.

10.3   Attenuated vaccines

DICTOL is a good example of an empirical vaccine. At the time of its development, and to a large extent even now, nothing was known about the nature of the relevant antigens or of the protective response they induce, but the vaccine was effective. Many similar irradiation-attenuated vaccines have been attempted since but, other than those against lungworms in sheep and goats, few have achieved comparable success. An irradiated larval vaccine was developed against hookworm infections (*Ancylostoma caninum*) in dogs, an important veterinary problem in the USA, and used commercially for some time. However, although this gave good protection against the pathological consequences of infection, it did not give complete resistance. The presence of hookworm larvae in the faeces of vaccinated dogs led to suspicions of vaccine failure, even though there was no overt disease, and the vaccine was eventually withdrawn in favour of anthelmintic treatment. Irradiation-attenuated larvae were also used as the basis of a vaccine against *Haemonchus contortus* in sheep. In experimental trials, using lambs of 7 to 8 months and adult sheep, the vaccine was highly effective, giving up to 98% protection. In the field, however, it proved difficult if not impossible to confer protection on young lambs, the age-group most at risk, or on older animals that had acquired early infections from pasture. This unresponsiveness of young lambs and the 'tolerance' of older infected animals remains unexplained, but are serious obstacles to successful vaccination.

These three examples illustrate some of the advantages and some of the problems associated with empirical development of vaccines and, in particular, with the use of attenuated parasites. Attenuated vaccines are useful, in that they solve some of the problems of availability of antigens and provide controlled exposure to living parasites, which often seems necessary for induction of an effective immunity. They are acceptable for use in animals, and a number of attenuated vaccines are available (Table 10.2). However, such an approach is not suitable for human vaccines, because of the risks of possible side-effects, vaccine failures, and incomplete attenuation.

Table 10.2 *Attenuated vaccines in use against parasites of domestic animals*

Infection	Nature of vaccine
Bovine schistosomiasis	Irradiated larvae
Bovine babesiosis	Passage-attenuated parasites
Ovine toxoplasmosis	Passage-attenuated parasites
Bovine theileriosis	Passage- and cell-line-attenuated parasites
Coccidiosis in poultry	Parasites with abbreviated life cycles

10.4 Modern anti-parasite vaccines

Vaccine research has been a major thrust of immunoparasitology during the last decade, and there have been many important developments in this field. The progress made can be considered under the five headings given above and then discussed in relation to specific infections.

10.4.1 Mechanisms involved in protective immune responses

There has been substantial progress in understanding the immunological interactions between hosts and parasites. In a number of infections (e.g. leishmaniasis, malaria, schistosomiasis, filariasis) the stages of the parasite that are vulnerable to immune attack have been identified, the effector mechanisms required for expression of immunity defined, and the control of these mechanisms analysed in detail. Equally importantly, it is now understood why immunity often appears ineffective or inefficient, and why infections may have immunopathological consequences. This knowledge makes it possible to think in terms of selectively boosting host-protective elements of the immune response, e.g. by targetted vaccination or by cytokine treatment rather than simply attempting overall stimulation of anti-parasite responses. The correct use of defined antigens and their appropriate presentation should make it possible to avoid those responses that are parasite protective (i.e. allow parasites to escape from immunity) and minimize undesirable pathological side effects.

10.4.2   Antigens that elicit protective immune responses

Progress in understanding and manipulating mechanisms of protective immunity is dependent upon progress in identifying the antigens involved in eliciting resistance. A number of technological advances have simplified the identification, isolation and characterization of these molecules. Among these are:

- *Separation techniques.* Polyacrylamide gel electrophoresis allows easy separation of parasite components, and can be combined with immunoblotting for identification of individual antigens. Preparative techniques based on iso-electric focusing or liquid chromatography make it possible to recover defined fractions of parasite components at high purity.

- *Monoclonal antibodies.* These allow specific recognition of single epitopes and have many uses. They can be used in the isolation of selected antigens from mixtures of molecules by a variety of preparative procedures (e.g. affinity chromatography), in screening recombinant expression systems to detect antigens of interest, and in the identification of target antigens in developmental stages of parasites.

- *Labelling techniques.* Antigens present in parasites, in their products, or in infected host cells and tissues, can be identified and characterized by labelling directly with isotopes, or indirectly with labelled antibodies or lectins, prior to recovery and separation.

- *T cell proliferation assays.* Responses of T cells to particular antigens or to components separated by electrophoretic techniques can be monitored directly by proliferation or by cytokine release.

- *Immunoepidemiological techniques.* Detailed surveys of populations exposed to infection, allied to analysis of serological responsiveness, make it possible to correlate particular antigen-recognition patterns with different infection/resistance categories.

The majority of these techniques have been applied to the identification of parasite molecules that are recognized during infection, on the premise that these are likely to provide good candidate vaccine antigens. It has been suggested, however, that if a parasite 'allows' the host to respond to an antigen, that antigen is unlikely to be important for parasite survival. An alternative strategy is therefore to use molecules not recognized during infection. One group of antigens that comes into this category of so-called 'concealed' ('cryptic' or 'hidden') antigens are those present on the intestinal surfaces of worms and arthropods. The host does not normally recognize these molecules, but they provide excellent targets for artificially induced responses (see below).

10.4.3   Production of antigenic material in quantity

Of the many antigens presented by living parasites to their hosts, relatively few are of major significance in stimulating protective immune responses. Therefore, even if large amounts of native parasite material can be obtained and the relevant antigens isolated, the yield of these antigens is likely to be small. For example, 1.2 kg of the tick *Boophilus microplus* is needed to prepare 100 mg of protective antigen – a recovery rate of 1 part in 12 millions! This difficulty has been one reason for considering attenuated parasites as a source of vaccine antigens, in addition to the fact that the living parasite is often much more immunogenic than antigens prepared from it. However, even though it is technically possible with some parasites to produce large enough numbers of infective stages for use in attenuated vaccines there are, as described above, a number of drawbacks to their use, in addition to problems such as limited life span and storage difficulties. For many other parasites this strategy is simply impossible, as adequate numbers of parasite stages cannot be produced.

Alternative approaches for parasite production include the use of *in vitro* techniques, and there have been some successes in this field, e.g. in the culture of *Plasmodium falciparum* and of bloodstream trypanosomes. However, although this technique has made available large numbers of organisms for antigen analysis and other immunological studies it cannot be considered as a realistic way of obtaining vaccine antigens. The alternative approaches therefore are those of recombinant DNA technology and chemical synthesis, both of which have become standard techniques in recent years.

Gene cloning allows the expression of eukaryote DNA in organisms such as *Escherichia coli* and yeast, or in insect cell lines, which can be cultured in bulk. Manipulation of the expression vector and expression system can allow large-scale production of peptide antigens and even of glycosylated molecules, which can then be tested for their immunological potential. The two basic approaches to cloning are to produce cDNA libraries, by transcribing the mRNA present in the target organism (i.e. message for molecules that are actually being produced), or genomic libraries from the entire genome. In both cases it is necessary to select out the recombinants that have the target gene or are producing the molecule of interest. The first can be achieved using labelled probes of a complementary nucleotide sequence, the second by screening with specific monoclonal antibodies or with polyclonal antibodies produced by infected and immune hosts. Alternative routes to obtaining relevant DNA include production of synthetic oligonucleotides based on known amino acid sequences, and production of longer sequences of DNA from smaller sequences by the use of the polymerase chain reaction.

When pure preparations of specific antigen peptides can be produced, either by recombinant means or by careful biochemical isolation, it is a relatively simple matter to determine the amino acid sequence. This can then be used to deduce the corresponding nucleotide sequence (for production of synthetic oligonucleotides) or used as a template for chemical synthesis of larger amounts of the peptide.

10.4.4   Stimulation of immunity under field conditions

Even when vaccines have been formulated there are many problems in their effective use under field conditions. One limitation is the requirement for vaccines to protect genetically heterogeneous hosts against genetically heterogeneous populations of parasite species. Early experiences with use of defined peptides from the circumsporozoite protein of *Plasmodium* in vaccines, in both mice and humans, showed that there was genetic restriction of the ability to recognize these peptides – some individuals lacking the class II MHC molecules necessary for antigen presentation and recognition. This particular problem can be overcome by using a cocktail of peptides, or by linking peptides to immunogenic carriers. Where there is extensive genetic heterogeneity in the target parasite population it is necessary to ensure that the vaccine antigens used are present in, or cross-react with, all strains likely to be encountered. Broader racial differences in vaccine responsiveness may also present a problem, as has been experienced in BCG vaccination campaigns. Two additional hazards face the use of anti-parasite vaccines in endemic areas. Firstly there is the need to vaccinate children or young animals early in life, before they are exposed to infection. It is well known that concurrent parasite infections can reduce the efficacy of vaccination and it is possible that transplacental transmission of antigen, or even parasites, from infected mothers may have the same effect. Secondly, vaccination may have to be carried out in populations that are less than optimally nourished. Undernutrition (both protein–calorie and trace element) is associated with immune depression and can exert marked effects upon the effectiveness of vaccination.

10.4.5   Commercial and user requirements

Commercial incentives for the development of vaccines against parasitic infections remain linked to the likelihood that there will be an economic return on the enormous investment necessary. In general, vaccines against infections prevalent in humans in developing nations offer little prospect for profitability, and their development and eventual use relies heavily upon support from the international funding agencies. The exceptions are those vaccines directed against prevention of parasitic diseases such as malaria in tourists and military personnel, and those against species responsible for opportunistic infections in developed countries. User requirements pose many problems in addition to the obvious ones of reliability and safety. There are considerable logistic difficulties in making labile vaccines available in tropical regions, where limitations of transport and cold storage facilities may restrict effective distribution. In addition there are the difficulties of persuading and organizing large numbers of people to attend, sometimes repeatedly, for vaccination, particularly if this involves injection. Development of vaccines that need to be given only once ('one-shot') or can be given orally would be a major step forward. Examples of the first are those that make use of recombinant vector organisms (e.g. vaccinia virus) to deliver antigens of a number of species, the second allow antigen to be administered by mouth, either via an attenuated recombinant vector such as *Salmonella* or in microparticles that allow controlled release of antigen. Many vaccines still have to be given with adjuvants if high-level responses are to be obtained, and this presents a further difficulty, as few adjuvants used in vaccine development are suitable for routine clinical use. The potential for engineering cytokine genes into recombinant vectors or attaching cytokine sequences to recombinant vaccines may provide solutions to this problem.

10.5   Progress in development of vaccines against specific infections

A large number of parasites are the subject of vaccine-related research at the present time. The progress made in relation to the use of attenuated live organisms in vaccines has been mentioned earlier (Table 10.2). In this section the focus will be on progress in developing molecular vaccines against some of the major diseases of humans and domestic animals. The steps involved in developing such vaccines to the point at which they can be used in the target populations are summarized in Table 10.3.

Table 10.3 *Stages in development of a molecular anti-parasite vaccine*

Demonstrate immunity against target species
Identify and characterize important antigens
Confirm immunogenicity of isolated/purified antigens
Clone and sequence antigen genes
Produce recombinant or synthetic antigens
Validate immunogenicity
Design vaccine and delivery system
Scale-up production
Test safety and efficacy in clinical/field trials
Obtain approval for commercial use
Market

10.5.1   Malaria

All stages in the cycle can be considered as targets for vaccination-induced immunity, but most attention has been focused on the sporozoite and the merozoite in terms of inducing responses that provide protective immunity for the host. In the case of *P. falciparum*, a great deal is now known of the antigens of these stages that can be used to elicit appropriate responses and a number of trials in primates and humans have been carried out.

Several vaccine studies have used recombinant or synthetic antigens derived from the sporozoite CS antigen. Early work graphically illustrated the problems of using relatively small peptide sequences as the basis of a vaccine. Not only were these sequences different between strains of *P. falciparum* in endemic areas, to which a given individual might be exposed, but recognition of their limited epitopes was subject to MHC restriction and a proportion of vaccinated individuals failed to respond. Several approaches can be adopted to overcome these limitations, e.g. use of immunogenic carriers, inclusion of CS-related sequences from different strains in a 'cocktail', but a fundamental objection to a vaccine directed only at the sporozoite is the consequence of vaccine failure. The sporozoite is exposed for a relatively short time to immune effectors and escape by only a small number (theoretically one) would be sufficient to initiate infection. Vaccines that generated responses against the exo-erythrocytic stages in the liver, for example by exploiting cytotoxic T cell responses against hepatocyte-expressed parasite antigens, would help to counteract this difficulty and progress is being made in this respect. However, even if vaccines directed at these stages prove effective, there is obvious merit

Table 10.4 *Vaccination trials against* P. falciparum *malaria using the SPf66 candidate antigen*

	Colombia 1991–2	Tanzania 1993–4
Type	Randomized, double-blind, placebo-controlled	Randomized, double-blind, placebo-controlled
Protocol	3 doses at 0, 4 & 20 weeks	3 doses at 0, 4 & 26 weeks
Assessment	*Malaria cases*:	*Clinical episodes (parasite density)*:
	Controls 29.9%	Controls 16.7% (3660)
	Vaccinated 20.6%	Vaccinated 12.0% (1716)
Efficacy	38.8% protection	31% protection

Source: (Data from Valero *et al.*, 1993, *Lancet*, **341**, 705 and Alonso *et al.*, 1994, *Lancet*, **344**, 1175.)

in designing a vaccine that would provide protection against both the pre-erythrocytic and erythrocytic stages, so that the parasite is controlled even if sporozoites escape.

Vaccines against the erythrocytic stages can be based upon antigens present at the surface of the parasitized red blood cell (PRBC), those released when the PRBC bursts, those on the merozoite surface, or all three. *In vitro* and *in vivo* data show that antibodies raised against antigens expressed on the PRBC surface are valuable in facilitating direct agglutination and phagocytosis. In addition, those directed against the molecules involved in sequestration prevent adherence to endothelial cells, thus helping not only to prevent pathology but also ensuring that PRBC circulate through the spleen, where they can be removed. Antigens released from ruptured PRBC have been used experimentally to provide protection against challenge with *P. falciparum* in *Aotus* and *Saimiri* monkeys. Antigens present on the merozoite surface have been the subject of the most intense study as vaccine candidates, and a number have been used in trials in primates and humans. For example, MSA-1 from *P. falciparum* has been tested successfully in *Aotus* and *Saimiri* monkeys, and some protection was also obtained in *Aotus* using a recombinant RESA protein. Some of the greatest interest has been focused on the antigen SPf66, a chemically synthesized peptide polymer developed by Pattaroyo *et al.* (1988). The polymer contains sequences from both sporozoite and merozoite proteins, and has been shown to protect *Aotus* monkeys and humans against challenge infection. It has given significant protection in an initial large-scale field trial and been used in a number of trials in Africa, Asia and South America (Table 10.4).

These approaches to a malaria vaccine are designed to protect the recipient against disease by enhancing immune responses to infection. Alternative strategies are to protect individuals against the severe clinical consequences of *falciparum* malaria, by immunizing against those parasite-derived molecules that initiate pathological responses (i.e. an anti-pathology rather than an anti-infection vaccine), and to protect populations by immunizing against antigens present on the sexual stages, thus preventing successful development after uptake by the mosquito (i.e. an anti-transmission vaccine). Neither approach has yet been the subject of trials, although a number of defined antigens are now available for these studies. As with all vaccines, malaria vaccines face the problem of enhancing responsiveness through the use of adjuvants. One way around this problem, which has been used experimentally in rodent models, is to engineer parasite antigens into an attenuated *Salmonella* vector, which can then be used to infect the host and provide continuous antigen release.

10.5.2   Schistosomiasis

Laboratory studies have clearly identified the early larval stages of infection as being both immunogenic and susceptible to immune attack. Irradiation-attenuated larvae have been used successfully to vaccinate experimentally infected rodents and primates, and have formed the basis of successful vaccine trials against *Schistosoma bovis* infections in cattle in the Sudan. The problems of supply of irradiated parasites has to some extent been overcome by development of successful protocols for cryopreservation but, for a variety of reasons, a similar vaccine cannot be considered as an option for use in humans. Not only is there the risk of incomplete attenuation, but experience with the bovine vaccine has shown that, while it protects well against the development of pathology, the immunity generated is not wholly effective against infection and vaccinated animals continue to develop egg-producing adults. Current approaches to development of vaccines suitable for human use have focused on the isolation and use of defined antigens. A number have been identified as vaccine candidates and some have been the subject of detailed trials in rodents and primates.

A major group of candidate antigens, which occur in all of the human schistosomes as well as in species such as *S. bovis*, are the glutathione S-transferases (GSTs), enzymes thought to protect against damage from oxygen radicals. There are two families of GSTs, with molecular weights of 26 and 28 kDa, and there is substantial sequence homology and serological cross-reactivity between GSTs within each family, regardless of the species of origin. It has

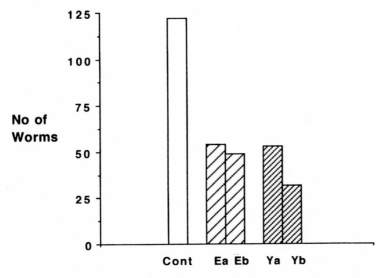

Fig. 10.2 Protection against infection with *Schistosoma mansoni* in rats immunized with recombinant Sm28GST antigen. Antigen was produced in *E. coli* (E) or in Yeast (Y) and given with either aluminium hydroxide (a) or BCG (b) as an adjuvant. Rats were infected with 1000 cercariae and worm numbers counted after 21 days. (Data from Gretzel *et al.*, 1993, *European Journal of Immunology*, **23**, 454.)

been shown in many experiments that immunization with these molecules confers substantial levels of protection, against both homologous and heterologous challenges, and accordingly they have been studied in great detail. Much of the work has been focused on the 28 kDa GST of *S. mansoni*, the gene for which has been cloned, sequenced and expressed. Purified, recombinant and synthetic peptides have been used successfully to protect rodents and primates against infection with *S. mansoni* and cattle against *S. bovis* (Fig. 10.2). Although vaccinated patas monkeys that were challenged with *S. haematobium* showed no reduction in worm numbers, worm fecundity was greatly reduced. Experimental studies in rodents have shown that IgE and IgA responses to GST are involved in eosinophil-mediated ADCC responses against the schistosomula stage, and that IgA may interact directly with adult worms with consequent effects upon their fecundity. In addition, immunoepidemiological studies of humans in endemic areas of Africa have correlated IgA responses to GST with development of resistance to *S. mansoni* infection.

The experimental studies of anti-schistosome vaccines have relied on injection of antigens with conventional adjuvants. Clearly, if such vaccines are to be a practical proposition for use in humans living in endemic areas alternative means of giving the vaccines must be developed. Several strategies have been tried for oral or nasal administration, including the use of attenuated *Salmonella typhimurium* as a vector.

10.5.3   Gastro-intestinal nematodes

Infections with intestinal nematodes are a major veterinary problem in a variety of domestic animals and particularly in sheep. In Australia alone, the cost to the sheep industry of nematode disease and control is more than US$500 millions each year. Despite repeated attempts, attenuated vaccines have been unsuccessful and control still relies on pasture management and frequent chemotherapy. Many of the most important species of nematode now show significant resistance to the commonly used anthelmintics, and effective immunoprophylaxis has become an urgent priority. Two strategies have been adopted in the search for effective molecular vaccines, one focuses on antigens that are recognized by infected hosts because they are released by the parasites, the other has concentrated on concealed antigens that are not normally available to the host. Although some success has been achieved experimentally with the first approach, e.g. in protecting sheep against *Trichostrongylus colubriformis* with purified or recombinant antigens, the most significant results have come from the second approach, particularly in protecting sheep against infection with *Haemonchus contortus*, which, because of its blood-sucking habit, is a major pathogen. Earlier ultrastructural studies of the worm's intestinal epithelium had identified an unusual polymeric protein, given the name contortin, which coats the surface microvilli. Sheep immunized against this protein showed enhanced resistance to challenge infection, which was clearly associated with antibodies directed against the intestine of the worm, thus identifying this as an important target organ. Other related epithelial molecules (glycoproteins designated H11 and H4.5) have been isolated and high levels of protection have been achieved after vaccination with H11, most significantly in merino sheep, which are very susceptible to this infection and express little resistance under normal conditions, and in young lambs, which are the necessary target population (Fig. 10.3). The degree of protection achieved was positively correlated with levels of serum antibody to H11. The genes for these molecules have been cloned, allowing recombinant material to be used in vaccine studies. H11 has aminopeptidase enzyme activity, and this activity is inhibited by serum antibodies from vaccinated animals. Related molecules have been been identified in other trichostrongyles of sheep, such as *Ostertagia circumcincta*, and these antigens have also been shown to confer protection.

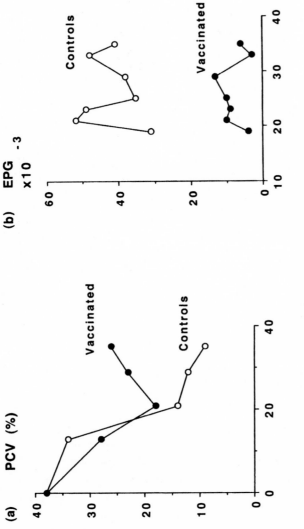

Fig. 10.3 Effect of prior vaccination with an H11-antigen of *Haemonchus contortus* on the pathological and parasitological consequences of infection in two-month-old Dorset Horn lambs. (a) Numbers of red blood cells, measured in terms of packed cell volume (PCV%); (b) course of infection measured by numbers of eggs per gram (EPG) of faeces. On day 35 the mean worm burdens of the two groups were: control lambs 7901 ± 788; vaccinated lambs 1331 ± 589. (Data from Taverner *et al.*, 1992, *Parasite Immunology*, **14**, 645.)

10.5.4   Larval tapeworms

The majority of mammalian tapeworms have life cycles that involve herbivorous intermediate hosts and carnivorous final hosts. Transmission occurs when intermediate hosts ingest eggs and when final hosts consume infected tissues containing viable larvae. Although humans act as final hosts for several species, the most serious diseases are those caused by accidental infections with the larvae of *Echinococcus* species (e.g. hydatid disease) and *Taenia solium* (cysticercosis). Larval tapeworms are also of considerable importance in cattle and sheep, which act as intermediate hosts for a variety of species. The presence in carcasses of larvae, whether or not these are harmful, is often sufficient for the meat to be condemned or fail to meet export requirements, and infections in sheep with dog-transmitted species such as *T. ovis* and *T. hydatigena* can be responsible for significant economic losses.

Detailed studies of experimental larval tapeworm infections in laboratory rodents and in rabbits have defined a number of important elements of the protective immune responses elicited by infection and by vaccination with parasite-derived antigens. Key points are that there is strong immunity (as there is in the field), that immunity acts against the early stages of larval development, that immunity is largely antibody-dependent, and that protective antigens are present in material released by cultured larvae. As with schistosome infections, there is concomitant immunity. Although animals given a primary infection may allow the larvae to develop to maturity, they are then resistant to reinfection. There are strong genetic influences on the ability to develop antibody responses and immunity to an initial infection, a slow response allowing larvae to establish and survive. However, both high- and low-responders express good resistance to reinfection, and both can be protected by vaccination. These laboratory studies provided realistic models of infection and immunity in the target hosts, and gave a firm rationale for the development of an effective vaccine.

Such a vaccine, using recombinant antigens, has been developed in Australia to protect sheep against infection with *T. ovis* and is now registered for commercial use. The vaccine is based upon antigens released by the oncosphere, the infective stage released from the egg, and its development followed the classical route of identifying molecules recognized selectively by immune sheep (47–52 kDa), making a cDNA library from hatched, activated onchospheres, cloning and expressing the genes in vectors, then using the fusion proteins to immunize animals. One particular molecule designated 45W was chosen as the major antigen and expressed in *E. coli* as GST-45B/X. Vaccinated sheep given two injections of 50 mg of antigen in saponin adjuvant expressed high levels of

Table 10.5 *Effects of vaccination with recombinant* Taenia ovis *antigen GST-45B/X on numbers of larval cysts in lambs exposed to field or experimental infection*

Group of lambs	Mode of infection	Number of cysts		% age protection
		Mean	Range	
Control	Field	92	36–348	–
Control	Experimental	156	74–254	–
Vaccinated	Field	2	0–11	98
Vaccinated	Experimental	12	0–203	92

Notes:
Field infections were acquired by grazing for 3 weeks on pasture contaminated by infected dogs. The experimental infection was 2000 *T ovis* eggs.
Source: (Data from Rickard, 1995, *Parasitology*, **110**, S5.)

immunity (> 90% resistance) to subsequent experimental challenge infection, and this protection lasted for about one year (Table 10.5). The importance of antibody-mediated mechanisms in this immunity means that immunized ewes are able to transfer significant protection to their lambs.

The vaccine does not cross-protect against other ovine tapeworms and gives no protection against *E. granulosus* or *T. saginata*. However, there is every possibility that similar vaccines could be developed against these species.

10.5.5 Ticks

The importance of cattle ticks as parasites in their own right and as transmitters of a variety of severe infections places a high priority on effective control. The intensive use of acaricides has led to resistance in tick populations and has emphasized the need for additional strategies, of which vaccination is one of the most important. Naturally exposed cattle do develop an immunity, which appears to be mediated largely through hypersensitivity responses to salivary antigens. However, vaccination with such antigens does not provide good protection. An alternative approach, used with the tropical cattle tick *Boophilus microplus*, has been to vaccinate against concealed antigens, using material from the midgut of the tick, and this has proved to be much more successful. Host antibodies raised against this material are taken in as the ticks feed, causing extensive damage to the intestinal cells. Cattle vaccinated in this way have shown substantial protection against challenge infesta-

Table 10.6 *Time scale of development of tick vaccine*

Vaccination trial using crude antigens	1982
Identification of Bm86 antigen	1986
First trial with recombinant antigen	1988
First field trial	1989
Provisional registration of vaccine	1992
Final registration of vaccine and first sales	1994

Source: (Based upon Willadsen, 1995 *Parasitology*, **110**, S43.)

tion in terms of numbers of ticks surviving on the body. Those ticks that do survive show reduced feeding and egg output. In combination these effects reduced total tick productivity over one generation by more than 90%.

The protective antigens in the midgut have been defined and isolated, and a glycoprotein designated Bm86 shown to be of major importance. The carbohydrate epitopes of this molecule appear to play no part in protection, this being a function of the peptide component. Recombinant antigen has been expressed in *E. coli* and insect cell lines, and used successfully to vaccinate cattle, giving effects on total tick productivity equivalent to those obtained with the crude antigen. A vaccine based on this antigen has been used in extensive field trials and is now registered for routine use. It is instructive to see the time course of the development of this vaccine (Table 10.6) in relation to the steps summarized in Table 10.3.

10.5.6   The future

The advances in approaches to anti-parasite vaccination summarized in this chapter have all been based on the presentation to hosts of parasite molecules, either derived from parasites themselves or produced as recombinant and synthetic molecules. An exciting alternative approach, which has already been used in model systems, e.g. in malaria, is the direct injection into hosts of DNA that codes directly for parasite antigens. Successful transfer of this genetic information into host cells allows production of the target antigens against which the host can then raise protective responses. Such short-circuiting of the steps involved in production and administration of vaccine material has many advantages, and may certainly prove an acceptable way of protecting animals against infection. Extension to use in humans, though feasible, may prove more problematic.

References and further reading

JOURNALS
Reviews and original papers on immunoparasitology appear in a wide range of
journals. Some of the major journals are listed below.

REVIEWS
Advances in Immunology
Advances in Parasitology
Contemporary Topics in Immunobiology
Current Opinion in Immunology
Current Opinion in Infectious Diseases
Current Topics in Microbiology and Immunology
Immunological Reviews
Immunology Today
Parasitology Today
Review of Infectious Diseases
Symposia of the British Society for Parasitology
Trends in Microbiology

PERIODICALS
Acta Tropica
American Journal of Tropical Medicine and Hygiene
Bulletin of the World Health Organization
Clinical and Experimental Immunology
European Journal of Immunology
Experimental Parasitology

Immunology
Immunology and Cell Biology
Infection and Immunity
International Journal for Parasitology
Journal of Experimental Medicine
Journal of Helminthology
Journal of Immunology
Journal of Infectious Diseases
Journal of Parasitology
Molecular and Biochemical Parasitology
Nature
Parasite Immunology
Parasitology
Parasitology Research
Proceedings of the National Academy of Sciences, USA
Research in Veterinary Science
Transactions of the Royal Society of Tropical Medicine and Hygiene
Tropenmedizin und Parasitologie
Veterinary Parasitology

TEXTBOOKS AND REVIEWS

The references listed below are intended to amplify the chapters in this book and to provide an introduction to the wider literature. With some exceptions original research papers have not been included as these can be obtained from the bibliographies in the references given. Texts on parasitology and immunology (Chapters 1 and 2) suitable for undergraduates on grounds of size and price are shown with asterisks.

GENERAL PARASITOLOGY

*Cox, F.E.G. (ed.) (1993). *Modern Parasitology* 2nd edition. Oxford: Blackwell Scientific Publications.

Cox, F.E.G., Kreier, J.P. & Wakelin, D. (eds.) *Parasitology*. Vol. 5 of *Topley & Wilson's Microbiology and Microbial Infections*. London: Arnold.

Englund, P.T. & Sher, A. (eds.) (1988). *The Biology of Parasitism*. New York: Alan R Liss.

*Hyde, J.E. (1990). *Molecular Parasitology*. Open University Press.

Mehlhorn, H. (ed.) (1988). *Parasitology in Focus*. Heidelberg: Springer-Verlag.

Mims, C.A., Dimmock, N., Nash, A. & Stephen, J. (1996). *Mims' Pathogenesis of Infectious Disease*, 4th edition. London: Academic Press.

*Muller, R. & Baker, J.R. (1990). *Medical Parasitology*. London: Gower Medical Publishing.

Schmidt, G.D. & Roberts, L.S. (1989). *Foundations of Parasitology* 4th edition. Baltimore: Williams & Wilkins.

*Smyth, J.D. (1994). *Introduction to Animal Parasitology* 3rd edition. Cambridge: Cambridge University Press.

Urquhart, G.M., Armour, J., Duncan, J.L. & Jennings, F.W. (1987). *Veterinary Parasitology*. Edinburgh and London: Churchill Livingstone.

Warren, K.S. & Mahmoud, A.A.F. (eds.) (1990). *Tropical and Geographical Medicine*. New York: McGraw-Hill.

GENERAL IMMUNOLOGY

Austin, J.M. & Wood, K.J. (1993). *Principles of Cellular and Molecular Immunology*. Oxford: Oxford University Press.

*Benjamin, E. & Leskowitz, S. (1996). *Immunology. A Short Course*. 3rd edition. New York: John Wiley.

Klein, J. (1990). *Immunology*. Oxford: Blackwell Scientific Publications.

*Kuby, J. (1997). *Immunology*, 3rd edition. New York: W.H. Freeman and Company.

*Janeway, C.A. & Traves, P. (1996). *Immunobiology. The Immune System in Health and Disease*, 2nd edition. Oxford: Blackwell Scientific Publications.

*Playfair, J.H.L. (1996). *Immunology at a Glance*, 6th edition. Oxford: Blackwell Scientific Publications.

*Roitt, I. (1997). *Essential Immunology* 9th edition. Oxford: Blackwell Scientific Publications.

*Roitt, I., Brostoff, J. & Male, D. (1998). *Immunology*, 5th edition, London: Mosby.

Romagnani, S. (ed.) (1997). *The Th1–Th2 Paradigm in Disease*. Heidelberg: Springer-Verlag.

IMMUNOPARASITOLOGY

Ash, C. & Gallagher, R. B. (eds.) (1991). *Immunoparasitology Today*. Cambridge: Elsevier Trends Journals.

Behnke, J.M. (ed.) (1990). *Parasite: Immunity and Pathology. The Consequences of Parasitic Infection in Mammals*. London: Taylor & Francis.

Immunological Reviews, vol. 112. (1989). Immunobiology of Parasites.

Immunological Reviews, vol. 127 (1992). Cytokines in Infectious Diseases.

Kaufmann, S.H.E. (ed.) (1990). *T Cell Paradigms in Parasitic and Bacterial Infections*. Heidelberg: Springer-Verlag.

Kierszenbaum, F. (ed.) (1991). *Parasitic Infections and the Immune System*. New York: Academic Press.

McAdam, K.W.P. (ed.) (1989). *New Strategies in Parasitology*. Edinburgh: Churchill Livingstone.

Sher, A. & Coffman, R.L. (1992). Regulation of immunity to parasites by T cells and T cell-derived cytokines. *Annual Reviews of Immunology*, **10**, 385–409.

Sher, A., Gazzinielli, R.T., Oswald, I.P., *et al.* (1992). Role of T-cell derived cytokines in the downregulation of immune responses in parasitic and retroviral infection. *Immunological Reviews*, **127**, 183–204.

Soulsby, E.J.L. (ed.) (1987). *Immune Responses in Parasitic Infections: Immunology, Immunopathology and Immunoprophylaxis.* Vol. 1 Nematodes, Vol. 2 Trematodes and Cestodes, Vol. 3. Protozoa. Boca Raton: CRC Press.

van der Ploeg, L.H.T., Cantor, C.C. & Vogel, H.J. (eds.) (1990). *Immune Recognition and Evasion: Molecular Aspects of Host–Parasite Interaction.* New York: Academic Press.

Wakelin, D. & Blackwell, J.M. (eds.) (1988). *Genetics of Resistance to Bacterial and Parasitic Infections.* London: Taylor & Francis.

Wang, C.C. (ed.) (1991). *Molecular and Immunological Aspects of Parasitism.* American Association for the Advancement of Science.

Warren, K.S (ed.) (1993) *Immunology and Molecular Biology of Parasitic Infections,* Oxford: Blackwell Scientific Publications.

Wyler, D.J. (ed.) (1990). *Modern Parasite Biology. Cellular, Immunological and Molecular Aspects.* New York: W.H. Freeman.

MALARIA

Berendt, A.R., Ferguson, D.J.P. & Newbold, C.I. (1990). Sequestration in *Plasmodium falciparum* malaria: sticky cells and sticky problems. *Parasitology Today*, **6**, 247–54.

Clark, I.A., Rockett, K.A. & Cowden, W.B. (1991). Proposed link between cytokines, nitric oxide and human cerebral malaria. *Parasitology Today*, **7**, 205–7

Hill, A., Elvin, J., Willis, A.C. *et al.* (1992). Molecular analysis of the association of HLA-B53 and resistance to severe malaria. *Nature*, **360** 434–39.

Kaslow, D.C. (1993). Transmission-blocking immunity against malaria and other vector-borne diseases. *Current Opinion in Immunology*, **5**, 557–65.

Kemp, D.J., Cowman, A.F. & Walliker, D. (1990). Genetic diversity in *Plasmodium falciparum. Advances in Parasitology*, **29**, 75–149.

Long, C.A. (1993). Immunity to blood stages of malaria. *Current Opinion in Immunology*, **5**, 548–56.

Melancon-Kaplan, J., Burns, J.M., Vaidya, A.B. *et al.* (1993). Malaria. In *Immunology and Molecular Biology of Parasitic Infections*, ed. K.S. Warren, pp. 302–51. Oxford: Blackwell Scientific Publications.

Playfair, J.H.L., Taverne, J., Bate, C.A.W. & de Souza, J.B. (1990). The malaria vaccine: anti-parasite or anti-disease? *Immunology Today*, **11**, 25–7.

LEISHMANIA

Alexander, J. & Russell, D.G. (1992). The interaction of *Leishmania* species with macrophages. *Advances in Parasitology*, **31**, 175–254.

Liew, F.Y. (1992) Induction, regulation and function of T-cell subsets in leishmaniasis. *Chemical Immunology*, **54**, 117–35.

Liew, F.Y. & O'Donnell, C.A. (1993). Immunology of leishmaniasis. *Advances in Parasitology*, **32**, 160–259.

Reiner, S.L. & Locksley, R.M. (1995). The regulation of immunity to *Leishmania major. Annual Review of Immunology*, **13**, 151–77.

Russell, D.G. & Talamas-Rohana, P. (1989). *Leishmania* and the macrophage: a marriage of inconvenience. *Immunology Today*, **10**, 328–33.

Sacks, D.L., Louis, J.A. & Wirth, D.F. (1993). Leishmaniasis. In *Immunology and Molecular Biology of Parasitic Infections*, ed. K.S Warren, pp. 237–68. Oxford: Blackwell Scientific Publications.

Scott, P. & Farrell, J.P. (1998). Experimental cutaneous leishmaniasis. Induction and regulation of T cells following infection of mice with *Leishmania major. Chemical Immunology*, **70**, 60–86.

TRYPANOSOMES

Albright, J.W. & Albright, J.F. (1991). Rodent trypanosomes: their conflict with the immune system of the host. *Parasitology Today*, **7**, 137–40.

Barry, J.D. & Turner, C.M.R. (1992). The dynamics of antigenic variation and growth of African trypanosomes. *Parasitology Today*, **7**, 207–11.

Cross, G.A.M. (1990). Cellular and genetic aspects of antigenic variation in trypanosomes. *Annual Reviews of Immunology*, **8**, 83–110.

Mansfield, J.M. (1990). Immunology of African trypanosomiasis. In *Modern Parasite Biology. Cellular, Immunological and Molecular Aspects*, ed D.J. Wyler, pp. 222–46. New York: W.H. Freeman.

Mansfield, J.M. (1994). T-cell responses to the trypanosome variant surface glycoprotein: a new paradigm? *Parasitology Today*, **10**, 267–70.

Vickerman, K., Myler, P.J. & Stuart, K.D. (1993). African trypanosomiasis. In *Immunology and Molecular Biology of Parasitic Infections*, ed. K.S Warren, pp. 170–212. Oxford: Blackwell Scientific Publications.

van der Ploeg, L.H.T., Gottesdiener, K. & Lee, M. G.-S. (1992). Antigenic variation in African trypanosomes. *Trends in Genetics*, **8**, 452–7.

HELMINTHS

Behnke, J.M., Barnard, C.J. & Wakelin, D. (1992). Understanding chronic nematode infections: evolutionary considerations, current hypotheses and the way forward. *International Journal for Parasitology*, **22**, 861–907.

Farthing, M.J.G., Keusch, G.T. & Wakelin, D. (eds.) (1995). *Enteric Infection: Mechanisms, Manifestations and Management, Vol. 2, Intestinal Helminths,* London: Chapman & Hall.

Kennedy, M.W. (ed.) (1991). *Parasitic Nematodes – Antigens, Membranes and Genes.* London: Taylor & Francis.

King, C.L. & Nutman, T.B. (1992) Biological role of helper T-cell subsets in helminth infections. *Chemical Immunology*, **54**, 136–65.

Maizels, R.M., Bundy, D.A.P., Selkirk, M.E. *et al.* (1993). Immunological modulation and evasion by helminth parasites in human populations. *Nature*, **365**, 797–805.

Lightowlers, M.W. & Rickard, M.D. (1988). Excretory-secretory products of helminth parasites: effects on host immune responses. *Parasitology*, **96**, S123–S166.

Moqbel, R. (ed.) (1992). *Allergy and Immunity to Helminths: Common Mechanisms or Divergent Pathways.* London: Taylor & Francis.

SCHISTOSOMES

Butterworth, A.E.B. (1994). Human immunity to schistosomes: some questions. *Parasitology Today*, **10**, 378–80.

Capron, A., Dessaint, J.P., Capron, M. & Pierce, R.J. (1992). Schistosomiasis: from effector and regulation mechanisms in rodents to vaccine strategies in humans. *Immunological Investigations*, **21**, 409–22.

Damian, R.T. (1989). Molecular mimicry: parasite evasion and host defence. *Current Topics in Microbiology and Immunology*, **145**, 101–15.

Hagan, P. & Wilkins, H.A. (1993). Concomitant immunity in schistosomiasis. *Parasitology Today*, **9**, 3–6.

Jordan, P., Webbe, G. & Sturrock, R.F. (Eds.) (1993). *Human Schistosomiasis.* Wallingford, Oxford: CAB International.

Newport, G.R. & Colley, D.G. (1993). Schistosomiasis. In *Immunology and Molecular Biology of Parasitic Infections*, ed. K.S Warren, pp. 387–437. Oxford: Blackwell Scientific Publications.

Smithers, S.R., Terry, R.J. & Hockley, D.J. (1969). Host antigens in schistosomiasis. *Proceedings of the Royal Society* B, **171**, 483–94.

Vignali, D.A.A., Bickle, Q.D. & Taylor, M.G. (1989). Immunity to *Schistosoma mansoni in vivo:* contradiction or clarification? *Immunology Today*, **10**, 410–16.

Wilson, R.A. & Coulson, P.S. (1989). Lung-phase immunity to schistosomes: a new perspective on an old problem. *Parasitology Today*, **5**, 274–8.

Wynn, T.A. & Cheever, A.W. (1995). Cytokine regulation of granuloma formation in schistosomiasis. *Current Opinion in Immunology*, **7**, 505–11.

GASTRO-INTESTINAL NEMATODES

Else, K.J. & Grencis, R.K. (1991). Helper T-cell subsets in mouse trichuriasis. *Parasitology Today*, **7**, 313–16.

Finkelman, F.D., Shea-Donohue, T., Goldhill, J., Sullivan, C.A., Morris, S.C., Madden, K.B., Gause, W.C. & Urban, J.F. (1997). Cytokine regulation of host defense against parasitic gastrointestinal nematodes: lessons from studies with rodent models. *Annual Review of Immunology* **15**, 505–33.

Reed, N.D. (1989). Function and regulation of mast cells in parasite infections. In *Mast Cell and Basophil Differentiation and Function in Health and Disease*, eds. S.J. Galli & K.F. Austen, pp. 92–110. New York: Raven Press.

Rothwell, T.L.W. (1989). Immune expulsion of parasitic nematodes from the alimentary tract. *International Journal for Parasitology*, **19**, 139–68.

Urban, J.F., Madden, K.B., Svetic, A. *et al.* (1992). The importance of TH2 cytokines in protective immunity to nematodes. *Immunological Reviews*, **127**, 205–20.

Wakelin, D., Harnett, W. & Parkhouse, R.M.E. (1993). Nematodes. In *Immunology and Molecular Biology of Parasitic Infections*, ed. K.S Warren, pp. 496–526. Oxford: Blackwell Scientific Publications.

FILARIASIS

Kazura, J.W., Nutman, T.B. & Greene, B.M. (1993). Filariasis. In *Immunology and Molecular Biology of Parasitic Infections*, ed. K.S Warren, pp. 473-495. Oxford: Blackwell Scientific Publications.

King, C.L. & Nutman, T.B. (1991). Regulation of the immune response in lymphatic filariasis and onchocerciasis. In *Immunoparasitology Today*, eds. C. Ash & R.B. Gallagher, pp. A54–A58. Cambridge: Elsevier Trends Journals.

Maizels, R.M. & Lawrence, R.A. (1991). Immunological tolerance: the key feature in human filariasis? *Parasitology Today*, **7**, 271–6.

Maizels, R.M., Sartono, E., Kurniavan, A. *et al.* (1995). T-cell activation and the balance of antibody isotypes in human lymphatic filariasis. *Parasitology Today*, **11**, 50–6.

Ottesen, E.A. (1992). Infection and disease in lymphatic filariasis: an immunological perspective. *Parasitology*, **104**, S71–79

ECTOPARASITES

de Castro, J.J. (1991). Resistance to ixodid ticks in cattle with an assessment of its role in tick control in Africa. In: *Breeding for Disease Resistance in Farm Animals*. eds. J.B. Owen & R.F.E. Axford, pp. 244–62. Wallingford, CAB International.

de Castro, J.J. & Newson, R.M. (1993). Host resistance in cattle tick control. *Parasitology Today*, **9**, 13–17.

Kaufman, W.R. (1989). Tick–host interaction: a synthesis of current concepts. *Parasitology Today*, **5**, 47–56.

Wikel, S.K. (1996). Host immunity to ticks. *Annual Review of Entomology*, **41**, 1–22.

VACCINATION

Bloom, B.R. (1989). Vaccines for the Third World. *Nature*, **342**, 115–20.

Bomford, R. (1989). Adjuvants for antiparasite vaccines. *Parasitology Today*, **5**, 41–6.

Capron, A., Dessaint, J.P., Capron, M. & Pierce, R.J. (1992). Vaccine strategies against schistosomiasis. *Immunobiology*, **184**, 282–94.

Good, M.F., Berzofsky, J.A. & Miller, L.H. (1988). The T cell response to the malaria circumsporozoite protein: an immunological approach to vaccine development. *Annual Review of Immunology*, **6**, 663–88.

Howard, R.J. & Pasloske, B.L. (1993). Target antigens for asexual malaria vaccine development. *Parasitology Today*, **9**, 369–72.

Immunology and Cell Biology, (1993) **71**(5). Vaccination against parasites.

Miller, T.A. (1978). Industrial development and field uses of the canine hookworm vaccine. *Advances in Parasitology*, **16**, 333–42.

Mitchell, G.F. (1989). Problems specific to parasite vaccines. *Parasitology*, **98**, S19–S28.

Morrison, W.I. (1989). Immunological control of ticks and tick-borne parasitic diseases of livestock. *Parasitology*, **98**, S69–S85.

Nardin, E. & Nussenzweig, R. (1993). T cell responses against preerythrocytic stages of malaria: role in protection and vaccine development against pre-erythrocytic stages. *Annual Review of Immunology*, **11**, 687–727.

Pattarroyo, M.E., Amador, R., Clavijo, P. *et al.* (1988). A synthetic vaccine protects humans against challenge with asexual blood stages of *Plasmodium falciparum* malaria. *Nature*, **332**, 158–61.

Phillips, R.S. (1992). Vaccination against malaria. *Immunobiology*, **184**, 240–62.

Rickard, M.D. (1989). A success in veterinary parasitology: cestode vaccines. In *New Strategies in Parasitology*, ed. K.P. McAdam, pp. 3–14. Edinburgh: Churchill Livingstone.

Selkirk, M.E., Maizels, R.M. & Yazdanbakhsh, M. (1992). Immunity and the prospects of vaccination against filariasis. *Immunobiology*, **184**, 263–81.

Sher, A. (1988). Vaccination against parasites: special problems imposed by the adaptation of parasitic organisms to the host immune response. In *The Biology of Parasitism*, eds. P.T. Englund & A. Sher, pp. 169–82. New York: Alan R. Liss, Inc.

Siimonds, R.S., Shearer, M.S. & Kennedy, R.C. (1997). DNA vaccines – from principle to practice. *Parasitology Today*, **13**, 328–31.

Smith, N. C. (1992) Concepts and strategies for antiparasite immunoprophylaxis and immunotherapy. *International Journal for Parasitology*, **22**, 1047–82

Targett, G.A.T. (ed.) (1991). *Malaria: Waiting for the Vaccine*. Chichester, Sussex: John Wiley.

Taylor, M.G., Hussein, M.F. & Harrison, R.A. (1990). Baboons, bovines and Bilharzia vaccines. In *Parasitic Worms, Zoonoses and Human Health in Africa*, eds. C.N.L. Macpherson & P.S. Craig, pp. 237–59. London: Unwin Hyman.

Tomley, F.M. & Taylor, D.W. (eds.) (1995). *Parasite Vaccines*. Symposium of the British Society for Parasitology Vol. 32. *Parasitology*, **110** (Supplement).

Valero, M.V., Amador, L.R., Galindo, C. *et al.* (1993). Vaccination with SPf66, a chemically synthesised vaccine, against *Plasmodium falciparum* malaria in Colombia. *Lancet*, **341**, 705–10.

Wakelin, D. (1994). Vaccines against intestinal helminths. In *Enteric Infection: Mechanisms, Manifestations and Management*, Vol. 2, *Intestinal Helminths*, eds. M.J.G. Farthing, G.T. Keusch & D. Wakelin, pp. 287–97. London: Chapman & Hall.

Index

Printed in the United States
18855LVS00005B/154-156